I0014129

Urcun Tanik

Architecting Automated Design Systems

Urcun Tanik

Architecting Automated Design Systems

VDM Verlag Dr. Müller

Imprint

Bibliographic information by the German National Library: The German National Library lists this publication at the German National Bibliography; detailed bibliographic information is available on the Internet at http://dnb.d-nb.de.

Cover image: www.purestockx.com

Publisher:
VDM Verlag Dr. Müller Aktiengesellschaft & Co. KG, Dudweiler Landstr. 125 a, 66123 Saarbrücken, Germany,
Phone +49 681 9100-698, Fax +49 681 9100-988,
Email: info@vdm-verlag.de

Produced in USA and UK by:
Lightning Source Inc., La Vergne, Tennessee, USA
Lightning Source UK Ltd., Milton Keynes, UK
BookSurge LLC, 5341 Dorchester Road, Suite 16, North Charleston, SC 29418, USA

ISBN: 978-3-639-00255-3

ARTIFICIAL INTELLIGENCE DESIGN FRAMEWORK
FOR OPTICAL BACKPLANE ENGINEERING

URCUN TANIK

ABSTRACT

The core contribution of this work is a new framework for architecture -driven

software engineering of large-scale, agent-based, complex systems for Knowledge-Based

Engineering (KBE). This reconfigurable and scalable framework fills a niche between

traditional requirements elicitation and design tool implementation. The new architectural

framework model, referred to as the Artificial Intelligence Design Framework (AIDF),

allows KBE system architects to achieve intellectual control over the high-level

development process. Therefore, structural modelers using object-oriented languages can

specify the coding requirements based on hierarchical decomposition of functional

requirements using axiomatic design principles. The engine block defined by the

framework has the capability to utilize networked knowledge repositories available

through intelligent agents acting on Web Services for the purpose of design risk

mitigation for reliability engineering.

KBE systems produced based on the new architecture framework automates the

design and inference processes for reliability engineering using an interlaced dual engine

block, developed during the National Aeronautics and Space Administration (NASA)

Fellowship. This type of intelligent automation of design support for product engineering

can save on development cost and time, while improving on quality. A comprehensive

solution is proposed to address this risk mitigation need for reliability engineering using a

System-of-Systems (SoS) approach, which consists of a synergistic overlap of many

broad topics such as design, agent modeling, and systems engineering. A case study implementation of the AIDF is developed using Acclaro Design for Six Sigma (DFSS) architectural development tool for configuring an optical backplane engineering application with design matrix, optimization, and verification techniques.

The Generic Architecture for Upgradeable Real-time Dependable Systems validation framework is introduced as a validation strategy for post-deployment expansion, after applying DFSS front-end validation during pre-deployment development. In addition to architecture validation, a comprehensive validation approach for a KBE SoS applications using the Synergistic Validation Methodology (SVM) for the AIDF has been developed. In conclusion, an AIDF-SVM is introduced as an architectural framework with a recommended validation methodology that functions as a platform for developing large-scale, reconfigurable and scalable KBE SoS applications.

DEDICATION

In memory of Dr. George Kozmetzky

ACKNOWLEDGMENTS

I would like to thank my chairman and committee members for allowing me to concentrate on a topic in computer engineering that I have always been keenly interested in developing expertise, while giving me the opportunity, providing me the resources, and equipping me with state-of-the-art knowledge and technology to be in a position to advance the art. I would like to express my sincere gratitude to Dr. Grimes for his kind guidance and mentorship of my dissertation as my chairman, and his invitation to contribute to the Alabama Telework Initiative. I appreciate his professional insight and extensive experience in business and academic aspects of innovative research, product development, and information systems. Furthermore, I would like to thank Dr. Vaughn and Dr. Joiner for serving on my committee and strengthening the real-world application of the abstract framework by emphasizing the importance of verification, validation, evaluation, and testing of complex systems interacting with Web Services. I would like to thank one of the founders of software engineering principles, Dr. Ramamoorthy of Berkeley, for also serving on my committee, listening carefully to my presentations, and giving his wholehearted blessing on my work whenever he visited UAB and at the Society for Design and Process Science (SDPS) conferences. I would like to thank Dr. Tanju, the professor of corporate accounting in the Department of Accounting and Information Systems, for providing his critical evaluation of the entire thesis from an information systems viewpoint that has significantly improved the presentation of difficult to explain engineering concepts for a global audience. Furthermore, as one of my

vi

esteemed mentors at UAB, I appreciate him for thoughtfully taking every opportunity to sharpen the precision of my thinking and approach. In addition to the committee members, I would especially like to thank Dr. Madni, founder and CEO of Intelligent Systems Technology, Inc., for his guidance and providing access to System-of-Systems (SoS) precursor technology during the proposal stage of my dissertation. I would like to thank Dr. Trevino of NASA Marshall Space Flight Center and leader in automated intelligence in rocket health-assessment research for providing critical background support, enthusiasm, and encouragement of my SoS dissertation application.

I am thankful to the School of Engineering for providing an excellent research environment, and a room with a view of the soccer fields next to the tennis courts. I would like to thank Dr. Jannett, as Graduate Program Director, for making sure that I excelled in supportive courses outside Computer Engineering, such as Artificial Intelligence (AI) and Probability in the Computer Science and Mathematics Department, respectively. I would like to thank Dr. Reilly from the Department of Computer and Information Science for allowing the flexibility to select my class project in AI, giving me the opportunity to experience HP Lab's Jena Framework and Java Expert System Shell from Sandia National Laboratories, implement Fuzzy Logic controllers, and survey decision support systems as part of my project presentation. I would like to thank Dr. Conner and Professor Green for giving me the opportunity to teach two lab courses in the Department of Electrical and Computer Engineering and Dr. Chernov of the Department of Mathematics for recommending me to teach mathematics. I would like to thank Dr. Smith, Dalton Nelson, Dr. Jett, and Dr. Callahan for their encouragement and supportive discussions in the hallways. I would like to thank those in the graduate school for their professionalism in providing administrative support as needed, such as Anita Moore and Curtis Jones, while working on the formatting process of my dissertation. I would like to thank the Engineering Librarian Craig Beard for always providing top-notch verification during my literature research, as needed. I would like to thank the engineering support staff, including Sandra Muhammad, Debbie Harris, Charlotte Oliver, and Maria Whitmire for assisting in my paperwork and providing a pleasant work environment.

I am grateful to the United States National Aeronautics and Space Administration (NASA) for awarding me a two-year NASA Training Grant Fellowship to develop and implement a prototype system based on my dissertation research. This critical support has allowed me to concentrate full-time on advancing my thesis to meet national standards of validation, while giving me the opportunity to consult with leaders in the Artificial Intelligence division at Marshall Space Flight Center. I appreciate Dr. Trevino and Jonathan Patterson for encouraging my studies in this field and exploring potential applications of my dissertation and ideas to advancing their work after graduation. I also appreciate many other accomplished NASA researchers, such as Bill Witherow and Tim Crumbley, who have welcomed and encouraged my research and presentations along the way. I appreciate the invitation by NASA to present my dissertation after graduation to receive further guidance on many diverse applications in both multidisciplinary and transdisciplinary information system fields, ranging from propulsion to enterprise engineering. I would like to thank Dr. Vohra, as NASA Campus Director, for supporting and guiding my NASA Fellowship and making it possible for me to succeed in a competitive environment for national grants by advising me well. I would like to thank Dr. Bangalore, professor in the Department of Computer and Information Systems, for bringing to my attention at the most critical time that an opportunity actually existed with NASA in the first place.

   I am thankful to the University of Alabama at Birmingham (UAB) for providing an outstanding environment for learning that enabled me to meet highly qualified faculty and students, while developing the groundwork for an exciting PhD topic with many applications in business and engineering, including but not limited to software and enterprise architecture. My very positive experience on campus also had a role in inspiring me to continue with my PhD at UAB to understand how to architect, develop, and manage large-scale, complex systems. As a result, thanks to Dr. Tanju's encouragement and sponsorship, I was able to found the Society for Entrepreneurship in Business and Engineering (SEBE) as a systems/enterprise engineering test-bed to help students reach their entrepreneurial goals by leveraging information systems. This experience provided an experimental venue in architecture-driven innovation and automated systems support for enterprise design. I would like to thank Dr. Gray from the Department of Computer and Information Science for providing an exemplary guest speaker program that enabled me to model that aspect of our SEBE service program. I appreciate the city of Birmingham, Alabama, for providing a supportive environment that helped me further assimilate important business ethics and values. Thanks to this solid foundation, I experienced personal, academic, as well as career growth over the past seven years, since my graduation from the University of Texas at Austin in electrical engineering and subsequent work experience with an international management consulting company and IC2 Institute of the University of Texas at Austin. I would like to thank Stan Gatchel and Michael Smith for hiring me as an information technology researcher and giving me real-world experience that motivated me to complete a Masters with thesis in Enterprise Engineering at UAB with a component in e-commerce. I would like to thank my corporate research director Louis Stoll, an outstanding career mentor, and Dr. Bedford, a resourceful manager, for their guidance and friendship that inspired me to launch a career in management information systems, while accommodating my telework arrangement and allowing summer vacation time for Euro-rail travel during World Cup'98 before starting full-time work.

I would like to thank the university president, Dr. Garrison, for fostering technology entrepreneurship at UAB, facilitating opportunity for all our SEBE members and allowing the financing of the kickoff party and brochures for our new organization. I appreciate the president and founder of TechBirmingham, Curtis Palmer, for enriching my UAB experience by serving as one of our guest speakers in SEBE, and allowing me to select ten outstanding students each year from our business group to represent UAB for two annual regional SevenCap Venture Capital conferences, in addition to listing our new organization in the Birmingham Reference Guide to Economic Development and Technology Organizations. I am thankful to Varadraj Prabhu Gurupur for facilitating SEBE's collaboration in the United States with foreign universities in Europe and Asia to reach several national newspapers of India. I am grateful to the UAB Business and Engineering School deans, Dr. Holmes and Dr. Lucas, respectively, for their critical background support and encouragement, allowing SEBE at UAB to thrive into a national society with international connections, as provided by its approved constitution by the UAB Leadership Office.

I would like to thank Dr. Szygenda of the United States Department of State, former engineering dean at SMU and UAB, for encouraging the growth of SEBE nationally, supporting the application of my research to enterprise engineering and entrepreneurship, and introducing my family to the beauty of Alabama. I would like to thank Dr. Sanabria and Dr. Sobrinho and families for their frequent visits from Spain and Brazil, respectively, encouraging my PhD work and entertaining collaboration of SEBE in the U.S with Europe and South America. I am appreciative of the dynamic talents of Daniel Sanabria, Bunyamin Ozaydin, Ozgur Aktunc, Rajani Sadasivam, and Cengiz Togay, who have been great friends and made my life in the lab enjoyable, while helping me to develop my skills in strategic management, as well as design. I would like to thank Ping Huang of China for volunteering to be the International Venture Capital Researcher of SEBE. I would like to thank Michael Lebeau and Mickey Gee of the Business School for serving as keynote guest speakers and connecting me to the local business community, which provided a springboard to found SEBE and architect its infrastructure.

I would like to thank Jonathan Barbee, Film and Television Producer/Small Business Owner and Entrepreneur, for volunteering to assist with media relations and serving as our first guest speaker, helping to kick-start our SEBE guest speaker program. Thanks to Paul Cleveland of SEBE USA and Damodar Prabhu of SEBE India, we have been able to develop a platform for joint collaboration between universities, advancing the large-scale development and management of the local UAB chapter and national SEBE Websites. I am especially thankful to Mickey Gee for finding me a position at the Office for the Advancement of Developing Industries (OADI) technology incubator to work as a technology researcher and selecting me, as part of a team, to represent UAB in two national student conferences, namely the Collegiate Entrepreneurship Organization (CEO) and National Collegiate Inventors and Innovators Alliance (NCIIA). I would like to thank Lori Woods and Melody Lake for inviting me to present SEBE to incoming MBA students during orientation. I would like to thank Dr. Singleton, Executive Director of Information Systems, for allowing his students to utilize SEBE as a resource to base a class project. I would like to thank David Anthony for providing me accurate insight from the trenches of how venture capital and investment is actually acquired in his course, "From Idea to IPO: the Technology Venture."

I would like to thank my friends and SEBE officers who have encouraged my PhD efforts: Ambassador Program Co-Developer Eisa Shunnara, Chief Information Officer Varadraj Prabhu Gurupur, Technology advisors Paul Cleveland, Damodar Prabhu, Rajani Sadasivam, and Ozgur Aktunc. I would like to thank SEBE Ambassadors: Adam Yost, Patrick Whitt, Andreas Wolf, Nancy Fuah, Richard D'Anna, Daniel Gallegly, Abhijeet Bhattacharya, Jessica Harmon, Daniel Sanabria, who also served as Secretary and our first Webmaster, Jonathan Torbert, who also served as Video Productions Officer, and Sajjad Hassan, who also served as BaseCamp software tools tester/Human Resources Manager. I would like to thank SEBE NASA outreach and service program members for their volunteer work, as needed, Baran Aksut, Anthony Geronimo, Sajjad Hassan, and Daniel Van Hausen. I would like to thank Praveen Gaure for volunteering his assistance with Endnotes. I would like to thank *Ingenuity* magazine and *Kaleidoscope* newspaper for reporting on our activities and honoring our members. I would like to thank Kazuhiro Abe for volunteering his language support from Japan. I also would like to thank Erhan Onal for composing entrepreneurial spirit theme music for SEBE.

I would like to thank my former advisors, mentors, and colleagues who have provided guidance and inspiration to pursue higher degrees, while developing my industrial experience. I appreciate Dr. Suh, Director of the Park Center for Complex Systems at Massachusetts Institute of Technology (MIT), and Dr. Ertas, Professor in the Department of Mechanical Engineering at Texas Tech University, for providing the groundwork for many design aspects of my dissertation. I appreciate everyone involved in the Society for Design and Process Science (SDPS) and the International Council for Systems Engineers (INCOSE) for either directly or indirectly assisting in the development of my career path. I am thankful to Dr. Kozmetzky for personally advising me on how to further develop my thesis and initiate a national organization to support entrepreneurs around the world with information systems, while connecting me with the right people to get started. I appreciate Dr. Gibson, Associate Director IC2 Institute, for encouraging me to follow Kozmetsky's ideas of sustained technopolis development and technology commercialization through virtual incubation in SEBE. I am thankful to Dr. Ronstadt, former Director of the IC2 Institute, for remembering my deliverable on university-based technology transfer and social entrepreneurship, while collaborating with Singapore on high-tech entrepreneurship, to request my service again, and engaging me with consulting work to help support myself right up to my NASA award.

I would like to thank Dr. Yeh, an important founder of software engineering, for inspiring me to develop a global company on a shoe-string budget in "Zero-Time" that has served as an example framework for new venture creation in SEBE by our entrepreneurial members. I am thankful to Dr. Sahinoglu, Professor in the Department of Computer Science at Troy University, who has enthusiastically recognized and promoted my entrepreneurial efforts. I would like to thank Dr. Watanabe and Dr. Yamaguchi of NEC Japan for their friendship and support at the SDPS conferences. I would like to thank Dr. Krämer, Chair of Data Processing Technology in the Department of Mathematics and Computer Science at FernUniversität in Hagen, Germany, and former president of SDPS, for setting an outstanding example of global technology and education leadership, as well as being a great friend of the family. I also would like to thank Dr. Messina, Chairman of the Department of Accounting and Information Systems, for his encouragements in the lobby throughout my dissertation research. I appreciate Alcatel Network Systems for the opportunity to experience full-time work as an undergraduate Co-Op intern right after a stint washing dishes at Arby's restaurant and training new members at President's Health Club, which gave me a newfound appreciation of free time that jolted me into action to understand the world of business and technology from a much more educated perspective.

I am grateful to all my family and friends who have provided moral support and encouragement in my five years of developing this dissertation. I would like to thank the Higginbotham family for all the camping trips as well as Thanksgiving and Christmas dinners, and especially lovely Janet for making me feel at home and inspiring me daily to work hard on my dissertation. She enhanced my student life with her excellent sense of humor and adventure, while motivating me with her sweet smile and understanding. I thank her for taking great care of our rambunctious little dog Chewy, while I was working late hours in the lab. I would like to thank the dynamic duo, Dr. Tuncer senior and Dr. Tuncer, M.D., for their enlightening humor. I would like to thank Dr. Bayrak, Dr. Kalan, Ulgar Dogru, Dr. Bugra Ertas, Alper Ertas, J.D., Dr. Jololian, Dr. Seker, Sambit Patnaik, Pinar Onay Durdu, Abidin Yildirim, Dr. Erbas, and Craig Owens for making me laugh every time I remembered their hilarious jokes, while writing my thesis late nights. I am thankful to the Ertas family, the Dogru family, the Aksut family, the Krämer family, the Tuncer family, the Gul Family, the Sobrinho family, the Sanabria family, the Oner family, the Owens family, the Bickel family, the Duchouquette family and all my family friends in Texas and beyond for providing a solid foundation to build my character. I would like to extend special thanks to my good friend Craig Owens for always treating me like a brother and sharing his artistic and intellectual insight like a modern day Leonardo da Vinci, since middle school. I would like to thank my childhood friend Dr. David Bickel and family for their friendship, and Dr. David Il Park of Korea, M.D., the valedictorian of our high school class, for giving me tough competition and setting a superb example of academic achievement that has inspired me to this day.

I would like to thank our good family friend Dr. Harris, Associate Dean at the University of Texas at Dallas and Executive Director at CyberSecurity and Emergency Preparedness Institute, for recommending me to President's Scholarship at SMU and instilling in me a profound sense of purpose right after high school graduation and acceptance to MIT. I would like to thank my good friend and our family doctor, Dr. Akhtar, M.D. for his care, advice, and enthusiastic support. I would like to thank my close family friends Dr. Dogru, Professor of Computer Science at METU, and Hakan and Dr. Fatos, Gul, M.D., for introducing me to the cultural richness of Turkish history, folklore, music, and dance. I appreciate Michael Lipscomb, J.D., for taking care of my sister while I worked on my thesis, his great sense of humor and particular selection of Hollywood blockbuster movies that have inspired me to develop SEBE and work with NASA.

I am eternally grateful to my close family, especially my mother Oya Tanik for making sure I did my homework and kept my promises and my father Dr. Murat M. Tanik for always advising me to do the right thing. I would like to thank my sister Yasemin Tanik for her care and affection, and taking me to my favorite concert the night before my PhD defense for inspiration. I thank my uncles Adem, Akil, Orhan, and my aunts Ayse, Sule, Nermin, Zerrin, Hulya, Meral, Ayten, and Hale, for encouraging my education and welcoming my visits with open arms. I would like to thank my cousin Tolga for visiting from Turkey, and showing me a great time at Disney World, right before I started my dissertation. I would like to thank my other cousins Orkun, Murat, Seniz, Gunay, Nilay, and Selay for also visiting and making me proud of their achievements. I am sure that, during these final days of dissertation writing and transition to a new chapter in my life, I may have forgotten to mention a number of very important people. I can only hope that they will forgive me for that.

Last, but not least, I appreciate my grandparents for passing on their values emphasizing the importance of higher education. With great respect and admiration, I thank my centurion grandfather Nurettin Tanik for still sharing with me his engineering insights and the ancient secrets of a long and happy life.

TABLE OF CONTENTS

Page

xxiii

LIST OF TABLES

LIST OF FIGURES

LIST OF ABBREVIATIONS

ADEB	AI Design Engine Block
ADT	Axiomatic Design Theory
ADTM	AI Design Task Manager
AIDF	Artificial Intelligence Design Framework
ANSI	American National Standards Institute
APG	Architecture Planning Group
ARS	Algorithm Reasoning Support
ATAM	Architecture Tradeoff Analysis Model
AWG	Architecture Working Group
BCM	Beam Combination Module
BSR	Board of Standards Review
CA	Customer Attributes
CAD	Computer-Aided Design
CBAM	Cost Benefit Analysis Method
CKAU	Centralized Knowledge Assimilation Unit
Cmap	Concept Map
CMMI	Integrated Capability Maturity Model
CODA	Cluster-on-demand architecture
CRMES	Center for Risk Management of Engineering Systems
CTDU	CommonKADS Task Determination Unit

CTS	Conant Transmission Support
CW	Continuous Wave
DAB	Data Allocation Bus
DE	Design Engine
DFSS	Design for Six Sigma
DHT	Design History Tool
DME	Device Modeling Environment
DMS	Data Mining Support
DMAIC	Define, Measure, Analyze, Improve, control
DP	Design Parameters
DR	Design Rationale
DRCS	Design Rationale Capture System
DRIM	Design Recommendation and Intent Model
DRL	Decision Representation Language
DRS	Domain Rule Support
DSA	Design Space Analysis
DSM	Design Structure Matrix
DTB	Data Transfer Bus
DVU	Data Validation Unit
EDS	Embodiment Design Stage
EOSDIS	Earth Observing System Data Information System
ESPRIT	European Strategic Program on Research in Information Technology
FLS	Fuzzy Logic Support

FMEA	Failure Mode and Effects Analysis
FR	Functional Requirements
FSOI	Free-Space Optical Interconnect
FTA	Fault Tree Analysis
GAS	Genetic Algorithm Support
GAURDS	Generic Architecture for Upgradeable Real-time Systems
GSFC	Goddard Space Flight Center
GUI	Graphical User Interface
HIDE	High Level Integrated Design Environment
HPKB	High Performance Knowledge Base
HOS	Hyper-Object Substrate
IBIS	Issue Based Information Systems
ICBM	Inter-Continental Ballistic Missiles
IDOV	Identify, Define, Optimize, and Verify
IEEE	Institute for Electrical and Electronics Engineers
INCOSE	International Council On Systems Engineering
ISS	International Space Station
JESS	Java Expert System Shell
KADS	Knowledge Analysis and Design Support
KAB	Knowledge Allocation Bus
KAE	Knowledge Assimilation Engine
KB	Knowledge Base
KBE	Knowledge-Based Engineering

KCE	Knowledge Correlation Engine
KJE	Knowledge Justification Engine
KTB	Knowledge Transfer Bus
KVU	Knowledge Validation Unit
MDA	Model-driven Architecture
MLAP	Machine-Learning Apprentice System
MLH	Hierarchical Multi-layer Design
MML	Moka Modeling Language
MSFC	Marshall Space Flight Center
MTTF	Mean time to Failure
MVC	Model View Controller
NASA	National Aeronautics and Space Administration
NNS	Neural Network Support
NRC	National Research Council
OBIT	Optical Backplane Interconnect Technology
OMG	Object Management Group
OO	Object-Oriented
OWL	Web Ontology Language
QFD	Quality Function Deployment
OPT	Optical Backplane Engineering
PIM	Platform-Independent Model
PLS	Predicate Logic Support
PRA	Probabilistic Risk Assessment

PSM	Problem Solving Method
PTTT	Process Technology Transfer Tool
PV	Process Variables
RAD	Rapid Application Development
RBD	Reliability Block Diagram
RCF	Rationale Construction Framework
ROI	Return On Investment
RUP	Rational Unified Process
RWN	Real-World Need
SADT	Structured Analysis and Design Technique
SATC	Software Assurance Technology Center
SE	Systems Engineering
SEI	Software Engineering Institute
SESC	Software Engineering Standards Committee
SoS	System-Of-Systems
SVM	Synergistic Validation Methodology
SWS	Semantic Web Service
SysML	Systems Modeling Language
TRF	Technology Risk Factor
TRIZ	Theory of Inventive Problem Solving
UML	Unified Modeling Language
XML	eXtensible Markup Language

CHAPTER I: OVERVIEW OF THE STUDY

INTRODUCTION

Managing large-scale, complex systems that experience greater risk with entan-gling dependencies and unforeseen interactions can have a deleterious impact on product engineering development cost, time, and quality [Muirhead, 1999; Muirhead and Simon, 1999; Muirhead, 2004]. Furthermore, as increasing system functionality for multi-disciplinary problems drives operational expansion, the validation and verification pro c-ess needs to be comprehensive and capable of adapting to evolving environmental condi-tions, fulfilling both end-user needs and engineering specifications alike, respectively.

An engineering need has been recognized that intellectual control over a Knowl-edge-Based Engineering (KBE) system assisting in the design process for product engi-neering composed of complex systems needs an overarching architectural framework. This type of approach provides a platform to structure the large-scale software engineer-ing development process for assuring ri sk mitigation during automated reliability engi-neering, with applications to National Aeronautics and Space Administration (NASA) [Trevino et al., 2005].

Although the current knowledge explosion on the Internet has created much o p-portunity for advances in KBE systems, keeping with these high-end performance de-mands leveraging Semantic Web Services, higher functionality of KBE systems have b e-come unmanageable without a new approach for intellectual control with new modeling techniques [Abdullah et al., 2001]. Hence, the introduction of supportive cross-disciplinary design solutions for product engineering that exploit knowledge repositories accessed by intelligent agents has come at a high price - unmanageable complexity directly impacting cost, time, and quality of KBE systems for many domains.

Consequently, a new paradigm of development of a KBE system-of-systems (SoS), spanning many fields is required, including systems engineering, design, agent-based modeling, and architecture. In this chapter, we will introduce our motivation, define our problem, summarize our solution, and outline our thesis approach.

MOTIVATION

A tighter integration of traditional requirements specification approaches and design tools has been recognized as a way to bridge the widening gap between requirements and implementation. In order to fill this niche, achieving comprehensive intellectual control over the development of today's multidisciplinary and complex engineering products has become an important goal. Both large-scale projects, such as the development of the International Space Station, and micro-scale projects, such as the development of optical backplanes [Broadbandrank, 2004], can benefit from this effort, thereby ensuring a broad impact on diverse fields. This need motivated the development of an architectural framework to serve as a platform providing a foundation to structure the systematic development of the engineering of such complex products. Furthermore, as increasing system functionality for multi-disciplinary problems drives operational expansion, the validation and verification process needs to be comprehensive and capable of adapting to evolving environmental conditions, fulfilling both end-user needs and engineering specifications alike, respectively. We were motivated to achieve intellectual control by developing a reconfigurable, scalable, and validated framework called the Artificial Intelligence Design Framework (AIDF) together with its Synergistic Validation Methodology (SVM).

Intellectual control over a KBE system assisting in the design process for product engineering composed of complex systems needs an overarching architectural framework to structure the large-scale software engineering development process for assuring risk mitigation during automated reliability engineering. A new agent-based approach, emphasizing risk mitigation, that is capable of continuously and systematically integrating global advances in expert systems, emphasizing artificial intelligence inference support, and KBE systems, emphasizing design support, for reliable product engineering is needed. Consequently, this dissertation was supported in part by a NASA Fellowship Training Grant to develop a reconfigurable, scalable, and validated architecture framework for a KBE system capable of managing large-scale, distributed, complex systems spanning multiple multi-disciplinary domains, including systems engineering, design, and agent-based modeling, and architecture. This type of system having constituent complex systems is called a System-of-Systems (SoS) problem, and the case study selected, optical backplane engineering design automation with intelligent agents, is such a problem requiring an architectural framework to achieve intellectual control over its complex, interacting SoS elements.

PROBLEMS WITH LARGE-SCALE AUTOMATED SUPPORT SYSTEMS

In the past, artificial intelligence had been heralded by some as a silver-bullet solution for many applications requiring a knowledge base, such as product engineering,
except that now the problem of automated design support has been compounded with too
many complex systems interacting at once [Russel and Norvig, 2003]. As the available
engineering knowledge skyrocketed [Kopena, 2003], in addition to evolving design algorithms and inference mechanisms, intellectual control over the increasing functionality of
the KBE system applied to design has suffered, as developers have been trying to keep up
with new demands on end-user functional requirements [Suh, 2001]. Previous KBE systems were limited only to specific domain applications and technologies and did not take
advantage of the benefits of model-driven architecture development. Furthermore, the
resulting applications generally were not designed to systematically leverage the advances in other fields, especially since many were developed at a time before the advent
of Semantic Web Services [Beckett, 2005]. Thus, frontier technology advances occurring
in other parts of the world, that could exponentially increase the power of the design support available during the design process, were not available or leveraged. As a result, the
concept of instantaneous support leveraging globally accessible resources was simply not
considered in their architecture. These new capabilities introduced to empower the designer include new technologies on the horizon arriving with Internet2 initiative [Inter-
net2, 2004] and the Next Generation Internet [NGI, 2006]. Already the field of intelligent agent technology is evolving fast to keep up with the knowledge explosion available
to designers using Semantic Web Services to identify and retrieve cross-disciplinary
component solutions for reliability engineering from remote knowledge repositories con-

tinuously updated by domain experts [Jett, 2006; Seker and Tanik, 2004]. New technologies are being developed to enable intelligent agents to interact with these authenticated knowledge depositories for component retrieval using early versions of the Web Ontology Language (OWL) [OWL, 2006; Kopena, 2003].

Our primary goal is to develop a state-of-the-art framework capable of achieving high-level intellectual control over object-oriented software engineering development of a KBE SoS application spanning multi-domain fields of engineering requiring design risk mitigation. We demonstrate the effectiveness of this AIDF solution with an application to free-space optical backplane engineering, i.e. Optical Backplane Interconnect Technology (OBIT), a suitable case study due to its SoS nature, on the micro-scale [Ayliffe, 1998].

Many of the hallmark features of an SoS application can be identified when trying to achieve intellectual control over the development of a KBE SoS application for OBIT, including systems engineering, agent-based modeling, design, and architecture elements. A KBE SoS application to OBIT, such as a Free-Space Optical Interconnect (FSOI) [Lee and Lee, 2006; Esener and Marchand, 2000; Kirk, 2003; Liu, 1997] , has system engineering issues such as assuring knowledge-based reliability engineering of its components in terms of availability, maintainability, and interface technology risks. The KBE SoS application has to deal with agent-based modeling concerns such as how to retrieve FSOI components from Web Services using intelligent agents and OWL. An KBE SoS application has design considerations such as FSOI optimal design configuration applying design rules and rationale based on domain expert feedback. Most importantly, from the perspective of the goal of this thesis, a KBE SoS application requires an overarching

architecture framework that can manage, properly structure, and integrate all the functional requirements of such a large-scale, multi-disciplinary, complex system that has
many overlapping and dynamic elements connected to Web Services.

Due to the high-precision requirements of OBIT, an emphasis on reliability engineering, a type of systems engineering, is ideal. Since there are a multitude of design
configurations possible for the same set of specifications, optimization of the design using rules and algorithms are involved. These forms of knowledge can be located locally,
as well as remotely, on networked knowledge repositories requiring all forms of systematic
and automated data mining techniques. Many of these techniques can be agent-
enabled and retrieve components that have already been modeled using OWL and deposited into these repositories continuously by scattered domain experts working independently. An overarching architecture framework is needed to integrate the multi-domain
elements into a coherent whole, so that the resulting KBE SoS application is developed
with design risk consideration in mind for reliability engineering.

FRAMEWORK FOR INTELLIGENT SYSTEM-OF-SYSTEMS AUTOMATION

An approach to managing large-scale, distributed complex systems interacting
with elements in multiple fields requires more than a systems engineering approach. Such
a new paradigm has emerged called System-of-Systems (SoS) that spans multiple fields,
including systems engineering, design, agent-based modeling, and architecture [Tanik, et
al., 2005]. Within systems engineering, systematic methods exist for design risk mitigation for reliability engineering that has been applied for product development. Many traditional design methodologies have been applied by government agencies notoriously

tackling SoS-type challenges, such as NASA, which heavily rely on systematic techniques, such as probability risk assessment, to identify system faults and vulnerabilities. As the Semantic Web emerged in the 21st century into a medium for intelligent agents to operate on remote, distributed, and authenticated knowledge repositories across the globe, new opportunities for leveraging Web Services for real-time, automated knowledge acquisition of modeled components became possible requiring agent-based modeling techniques.

Former KBE systems that did not account for the latest advances in the Web, including the Internet2 initiative and Next Generation Internet, rapidly became antiquated, without addressing the need to evolve to meet current demand for flexibility in configuration that leverages such Semantic Web Services and intelligent agents [Kopena, 2003]. Most importantly, a need was recognized by NASA that intellectual control over a KBE system assisting in the design process for product engineering composed of complex systems needs an overarching architectural framework to structure the large-scale software engineering development process of a KBE SoS to provide flexibility.

In this dissertation, we have worked with NASA Marshall Space Flight center over two years in developing such an architectural framework to achieve intellectual control over increasing software functionality interacting with complex systems in multiple domains. The resulting reconfigurable, scalable, and validated framework is called the Artificial Intelligence Design Framework (AIDF), which is capable of leveraging the potential of Web Services and networked knowledge repositories interacting with hundreds of domain experts and intelligent agents. The AIDF provides a platform to configure a series of KBE SoS applications based on a validated engine block design having twenty modular functions interacting with Web Services to support product design risk mitiga-tion. The AIDF assimilates local and global knowledge for processing, correlates and prepares this knowledge for analysis, and justifies the design rationale for output display of the final design recommendations in real-time during operation of the three engine blocks: (1) Knowledge Assimilation Engine (KAE), (2) Knowledge Correlation Engine (KCE), and (3) Knowledge Justification Engine (KJE). Hence, the KAE block is involved in assembling multiple forms of knowledge in the form of rules and ontologies for proc-essing in the KAE block using the design and inference engine, so that the model-view-controller can display the results to the designer. The dual engine block having 11 mod-ules in the design engine and 9 modules in the inference engine can be hierarchically de-composed using Acclaro Design for Six Sigma (DFSS) [DFSS, 2006], a high-level archi-tectural development tool recently acquired by General Dynamics, for detailed specifica-tion [Axiomatic Design, 2005; El-Haik, 2004]. Furthermore, we can configure the framework so that the designated architectural components can be made to interact with each other, as well as a set of intelligent agents, depending on the needs of the applica-tion. Furthermore, we show how the AIDF engine blocks can be expanded to include more modules using a design matrix that can expand the set of architecture components for functionality, using appropriate design parameters matching each functional require-ment during hierarchical decomposition. This axiomatic approach to system design is important for verification and validation by meeting the engineering specifications and end-user needs, respectively, by exposing and resolving any entangling dependencies. Finally, we show that the AIDF engines support all three design phases, namely conceptual, embodiment, and detail design, during reliability engineering, including designation of the active engine modules for each stage.

A Synergistic Validation Methodology (SVM) is introduced for comprehensive pre-deployment and post deployment AIDF validation based on the Design for Six Sigma [Axiomatic Design, 2005] and Generic Architecture for Upgradeable Real-time Dependable Systems (GAURDS) validation framework [Bondavelli, 2001]. The AIDF-SVM is a three-layer stratification approach for validating four target divisions (1) Software architecture, (2) Design Process, (3) Artificial Intelligence, (4) Knowledge acquisition. Each division is categorized into multiple areas, which are all further broken down into methodologies in terms of standards and techniques that can be used for continuous validation of the AIDF over time.

The broad impact of this thesis can be felt in those engineering disciplines that require cross-fertilization of engineering concepts in an integrated framework that provides a way to systematically configure KBE SoS applications suitable for each field.

OUTLINE

- *Chapter I*: In the overview of study chapter, we introduce the thesis motivation, the problem definition, our approach to solution, and outline.

- *Chapter II*: In the background and state-of-art chapter, we provide a literature review for a reconfigurable architectural framework application developed for reliability engineering automation leveraging KBE techniques networked to domain experts using intelligent agents. Broad topics integrated by such a SoS application include many interrelated fields, such as systems engineering, software and product engineering design, software and system architecture development, risk mitigation, reliability engineering, intelligent agents, human factors engineering, and knowledge-based engineering utilizing quality design and artificial intelligence algorithmic techniques. An in-depth review of these topics provide a foundation for intelligently automating the engineering design process for reliable product design impacting the fields of computer engineering, system-of-systems, and architecture development. We will show that intellectual control of a KBE system-of-systems, characterized by a large-scale system capable of managing a set of complex, distributed systems, can be achieved by adapting a reconfigurable and scalable architectural framework for reliability engineering automation.

- *Chapter III*: In the Artificial Intelligence Design Framework (AIDF) chapter, we introduce and define the AIDF, providing its scope and impact. We provide a validation overview for the architecture framework. We provide a discussion of the chapter contents, before describing the AIDF input, processing, and output block that operate as the Knowledge Assimilation Engine (KAE), Knowledge Correlation Engine (KCE), and the Knowledge Justification Engine (KJE), respectively. We describe each engine block in detail, providing snapshots of the implementation with Acclaro Design for Six Sigma. We describe the operations in each block, followed by an introduction of the three stages of the AIDF corresponding to the three design phases. The modular engine operations in each of the three AIDF design stages are examined with a design matrix. A discussion of the strategy used for development of the architecture is provided. The vendors supplying technology for each of the twenty modules in operation during reliability engineering automation are described in Appendix B.

- *Chapter IV*: In the case study chapter, we introduce the application, configuration, implementation, optimization, and verification, using the AIDF architecture framework for Optical Backplane Engineering [Robertson et al., 2000] . In the application section, we describe the Free-Space Optical Interconnect (FSOI) components and how to apply knowledge engineering [Studer et al., 1998] allowing domain experts to capture their tacit and explicit knowledge on components and design rationale, using methods such as web-enabled DSM templates, rules, and ontologies. In the configuration section, we show various levels of granularity in analyzing the AIDF, focusing at high-level down to engine module and element interaction analysis. In the implementation section, we provide screenshots using Acclaro DFSS design matrix architecture development tool. In the optimization and verification section, we show how the DFSS approach combined with the GAURDS validation framework, provides a pre-deployment and post-deployment validation strategy for the AIDF architectural framework

- *Chapter V*: In the synergistic validation chapter, we provide a methodology for validation for the architecture and the complex systems of a KBE System-of-Systems (SoS) application. In this chapter, we introduce the standard approach to engineering validation, followed by introducing the Synergistic Validation

Methodology (SVM) we used as our comprehensive approach to validation for the AIDF. This AIDF-SVM is shown to be divided into four divisions addressing multiple areas that correspond to well-defined methodologies in terms of standards, techniques, and methods employed today. We provide a rational for each division and area of validation. Supportive validation research work, such as terminological surveys, best practices for decision support were conducted, as well as peer-reviewed development work with NASA Marshall Space Flight Center over two years. We concentrate our validation on the software architecture division, which forms the bedrock for development of the AIDF.

- *Chapter VI*: In the conclusion, contributions, and future work chapter, we introduce the contributions of the AIDF in terms of broad impact of the framework on design risk mitigation for reliability engineering in the optical backplane domain, as well in other fields impacted by architecture development, system-of-systems, and computer engineering. Future work discussing how to systematically expand the architectural framework scope and its possible impact on commercialization efforts in optical backplane engineering is introduced.

- The *Appendices* provide details on the relevant details for the validation tables, coding, tests, and various other surveys conducted.

- The *List of References* provide a list of citations for each chapter.

CHAPTER II: BACKGROUND AND STATE-OF-THE-ART

OVERVIEW

In this chapter, we will provide a literature review for a reconfigurable architectural framework application developed for reliability engineering automation leveraging

Knowledge-Based Engineering (KBE) techniques networked to domain experts using

intelligent agents. Broad topics integrated by such a SoS application include many interrelated fields, such as systems engineering, software and product engineering design,

software and system architecture development, risk mitigation, reliability engineering,

intelligent agents, human factors engineering, and KBE utilizing quality design and artificial intelligence algorithmic techniques. An in-depth review of these topics provide a

foundation for intelligently automating the engineering design process for reliable product design impacting the fields of computer engineering, system-of-systems, and architecture development. We will show that intellectual control of a KBE system-of-systems,

characterized by a large-scale system capable of managing a set of complex, distributed

systems, can be achieved by adapting a reconfigurable and scalable architectural framework for reliability engineering automation.

MOTIVATION FOR INTELLECTUAL CONTROL OVER KBE SYSTEM-OF-SYSTEMS

Automating reliable product design leveraging networked knowledge sources requires is an architectural development challenge within the realm of a System-of-Systems

(SoS) paradigm. Identifying this thesis problem as a KBE-SoS is primarily due to the

large-scale complex system interaction with networked domain sources of a class of KBE systems specialized for design support. In order to fuse the operations of many fields together into a coherent whole, many disparate, yet interrelated areas had to be woven together into synergistic operation through our research. The topics in this chapter are meant to be broad introductions for deep subjects, which are delved in more detail in later chapters and the appendices. The sections in this chapter are separated into the following conceptual blocks (Table 2.1). The architecture framework developed in this thesis is intended to provide *intellectual control* over the constituent complex systems defined by these broad topics for the purpose of design risk mitigation for reliability engineering.

Table 2.1. Chapter sections addressing system-of-systems KBE problem.

Chapter Sections	Motivation
SYSTEMS ENGINEERING APPROACH FOR PRODUCT DESIGN	*Introduce design process, systems engineering, scope out architecture and need for intellectual control over KBE systems and briefly describe how it can be implemented and validated*
ARCHITECTURE RISK MITIGATION AND RELIABILITY ENGINEERING	*Introduce architectural validation, reliability engineering at component level, risk mitigation for system level*
KBE SYSTEM-OF-SYSTEMS APPROACH FOR DESIGN RISK MITIGATION	*Introduce KBE system-of-systems approach to automated design risk mitigation for reliability engineering*
VALIDATION APPROACH FOR KBE-SOS	*Show overall list of validation methods linked to Appendix A.*

SYSTEMS ENGINEERING APPROACH FOR RELIABLE PRODUCT DESIGN

A system is a collection of interconnected parts, which can be reliably engineered for design risk mitigation through reliability engineering, a form of systems engineering [Maier and Rechtin, 2000]. Before we define reliability engineering, we will explain systems engineering.

PRODUCT ENGINEERING DESIGN PHASES

Product Engineering Design Phases

Product engineering can be divided into three phases: *Conceptual Design, Embodiment Design,* and *Detail Design* [Pahl and Beitz, 1988] (Table 2.2). Proceeding through three phases, which are sequentially validated, final design blueprints can be produced (Fig. 2.1). Throughout the product engineering design phases, systems engineering provides all types of design risk mitigation support for reliability engineering.

Table 2.2. Design phases during product engineering.

Design Phases	Description
Conceptual Design Phase	Identifies essential problems, establish function structures, search for solution principles, combine and firm up concept variants.
Embodiment Design Phase	Determines preliminary layouts and configurations, selecting the most desirable preliminary layouts and refining and evaluating against technical and economic criteria.
Detail Design Phase	Optimizes configurations, shapes, weights and interactions by dimensioning and tolerance calculations

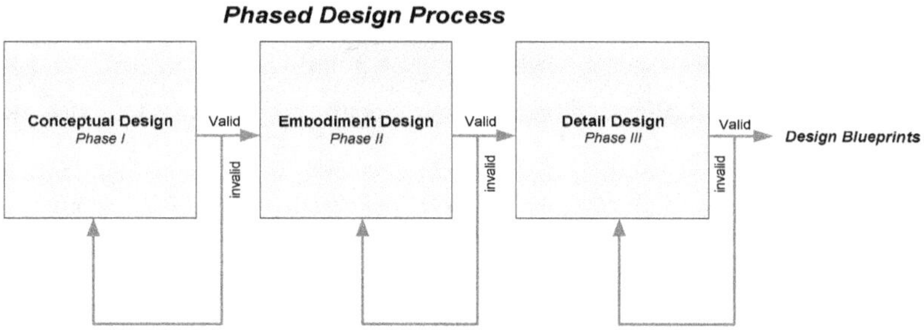

Phased Design Process

Fig. 2.1. Phased design phases producing blue-prints.

Hierarchical Decomposition of a System for Analysis and Development

Analysis of a product design entails an initial breakdown of the structure into its

constituent parts (Fig. 2.2), so that reliability engineering can be applied to its comp o-

nents. An approach of analysis is hierarchical decomposition where a product's systems,

subsystems, and components are identified in order to determine their individual intera c-

tions [Clarkson, 2001].   Suh's axiomatic design approach for systematic requirement

specification defines the functional requirements and design parameters for a design

based on the hierarchical decomposition process [Suh, 2001].

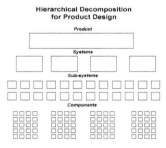

Fig. 2.2. Hierarchical decomposition process for product development.

SYSTEMS ENGINEERING APPROACH TO PRODUCT DEVELOPMENT

Defining Systems Engineering

An interdisciplinary approach for enabling the realization and deployment of

successful systems, Systems Engineering (SE) can be considered a type of holistic

systems thinking which enables a general application of engineering *technique*s to the

design process for all types of varied systems needing integration [Thome, 1993].

Characterizing systems and subsystems, and clarifying the interactions among the components and component clusters by applying various methods of risk management is a primary goal for systems engineering. Generally, systems engineering techniques involves the application of abstract paradigms to specific engineering disciplines involved in product engineering [Larsen and Buede, 2002]. The field of Systems Engineering supports the design process by performing the following (Table 2.3). System engineering considerations also include many aspects and views of the problem that require requirements and results documentation, design synthesis and analysis, as well as system validation at various stages of development.

Table 2.3. Systems engineering supports design process in these areas.

Area of Design Support	Description
Requirement Specifications	Defining and managing system requirements
Risk Mitigation	Identifying and minimizing risk
Component Integration	Integrating system components
Design Complexity Management	Managing system complexity
Coherent systems understanding	Enhancing communication and system understanding,
Informed decision-making	Knowledge application based on industrial best practices
Validation	Verifying that products and services meet customer needs

Defining the Roles of Systems Engineers in the Design Process

In this thesis, we emphasize the needs of a *system designer* from a product engineering point of view that provides risk management through application of reliability techniques. Peripheral roles that support this role include systems designer, systems analyst, validation/verification engineer, process engineer, as well as the systems engineer responsible for integration between subsystems. However, systems engineering roles are multiple, varied, and sometimes overlapping, based on a research conducted by INCOSE during its inception [Bielawski and Lewand, 1991; INCOSE, 2006a]. We provide a total of twelve different categories based on this research published in the in the inaugural issue of Systems Engineering (Table 2.4). Systems engineers are involved in the production of a product from conception to operation, as well as providing extended maintenance after rollout and deployment.

Table 2.4. Systems engineering roles.

System Engineering Role
Requirements Owner
System Designer
System Analyst
Validation/Verification Engineering
Logistics/Ops Engineer
Glue Among Subsystems
Customer Interface
Technical Manager
Information Manager
Process Engineer
Coordinator
Classified Ads SE

ARCHITECTURE FRAMEWORK DEVELOPMENT FOR SYSTEMS ENGINEERING

Defining Architecture Scope in Systems Engineering

The processes associated of the Tufts' Systems Engineering Process Model product lifecycle model are based on the Integrated Capability Maturity Model (CMMI) [INCOSE 2006a]. It should be noted that, with respect to the scope of the thesis, we are focusing on architecture development and validation for a KBE system, which actually comprises only one of eight distinct aspects of the systems engineering product life cycle

(Fig 2.3).   In this case, we are showing that the high-level architecture developed in this thesis can develop configured KBE systems as a "product" line. The architecture in this diagram can also refer to that developed for the structure of components by engineers, a process itself that can be automated.

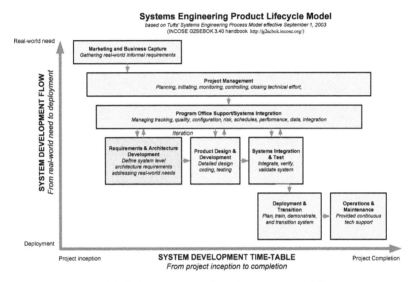

Fig. 2.3. Tufts' systems engineering process model.

SOFTWARE ARCHITECTURE STANDARDS HISTORY

Software architecture has developed over the last decade with attempts towards standardization of terminology (Table 2.5). The *First International Workshop on IT Architectures for Software Systems* was held in 1995 [Garlan, 1995a; Garland 1995b]. Although there have been many different definitions of architecture, this landmark conference resulted in a better definition of the various approaches taken in software architecture construction. Shortly following this workshop, the U.S. Congress passed omnibus

legislation in 1996 aimed at standardizing the definition for the nation [Filman, 1998;
Robinson and Gout, 2004]. An important development in IT architecture practice has
been the emergence of standards for architecture description, as expressed in ANSI/IEEE
Std. 1471-2000. The ANSI Board of Standards Review (BSR) approved IEEE's Recommended Practice for Architectural Description of Software-Intensive Systems as an
American National Standard in 2001 and Object Management Group (OMG) adopted the
Model-Driven Architecture approach in the same year. Both ANSI/IEEE Std. 1471-2000
and MDA are continuously being updated over the years, as architectural definitions and
standards evolve. In addition to the five architectural models, various styles were identified: *Pipe and filter style*, *OO style*, *event-based style*, *layered style*, *blackboard style*.

Table 2.5. Historical architectural development into a discipline.

Title	Date	Source	Architecture Clarifications
Architecture for KBE System-of-Systems	2006	Intelligent Systems, Inc. (Azad)	Evolving definition: Architecture for a large-scale system-of-systems project comprised of many constituent complex, distributed systems in fields such as design, systems engineering, and agent-based modeling leveraging Web Services and Semantic Web
Model-Driven Architecture	2001	Object Management Group (OMG)	Functional approach to architecture: Based on a platform-independent model (PIM) plus one or more platform-specific models (PSM) and sets of interface definitions, MDA development focuses first on the functionality and behavior of a distributed application or system, thereby divorcing implementation details from functionality
International Council for Systems Engineers (INCOSE) standard	2000	INCOSE	Definition: The fundamental and unifying system structure defined in terms of system elements, interfaces, processes, constraints, and behaviors
ANSI/IEEE Std. 1471-2000	2000	IEEE	AWG Definition: *The fundamental organization of a system, embodied in its components, their relationships to each other and the environment, and the principles governing its design and evolution.* (Recommended Practice for Architectural Description of Software-Intensive Systems)
National Defense Authorisation Act for Fiscal Year 1996 (Clinger-Cohen Act)	1996	U.S. Congress	Legal Definition: Information Technology Management Reform Act of 1996 *An integrated framework for evolving or maintaining existing technology and acquiring new information technology to achieve the agency's strategic goals and information resource management.*
First International Workshop on IT Architectures for Software Systems	1995	International IT Workshop	Categorization: 5 architectural models categorized and defined by Shaw: *Process Model, Framework Model, Structural Model, Dynamic Model, and Functional Model.*

SOFTWARE ARCHITECT ROLE

Emergence of Software Architecture as a Distinct Discipline

Software architecting is emerging as a discipline in itself, independent of software engineering [Maier and Rechtin, 2000]. Systems architecture has been identified as one of the four fundamental disciplines for engineering systems: (1) systems architecture/systems engineering and product development, (2) operations research and systems analysis, (3) engineering management and (4) technology and policy [Hastings, 2004]. Based on this analysis, we are focusing on integration of the concepts in the first set of disciplines, comprised of systems architecture/systems engineering/product development, by examining the role of the architect tackling a *system-of-systems* problem composed of complex constituent systems interacting with Web Services. Within this particular set, architecture itself is emerging as a discipline [Shaw, 1996]. The other three fundamental disciplines mentioned by Hastings for engineering systems are outside the scope of this thesis. By providing technical design, the architect must translate the end-user requirements in the real-world domain into a language that software developers and systems engineers understand for detailed design, coding, and testing (Fig. 2.4) [Jayaswal and Patton, 2006]. Hence, the steps are (1) requirements analysis by domain experts, (2) functional specification by system analysts, (3) technical design by software architects, (4) detailed design by software engineers, (5) coding by programmers, and (6) testing by quality assurance. Thus, the architect must play the role of mediator using a common language to bridge the gap between the customer and the actual engineers directly involved in any given project. This common language is enhanced by modeling with views using methods such as aggregation, partitioning, and certification.

1 Requirements Analysis by Domain Experts	2 Functional Specification by System Analysis
3 Technical Design by Software Architects	4 Detailed Design by Software Engineers
5 Coding by Programmers	5 Testing by Quality Assurance

Fig 2.4. Traditional software development processes.

Model-Driven Architecture Development of Technology Independent Platform

Architecture development at a higher level of abstraction emphasizing late-binding of software technology has become very popular for current architectural development. Emerging standards of the Object Management Group (OMG) such as model-driven architecture (MDA) [OMG, 2006] are providing guidelines on how to ensure development of a technology independent platform. The MDA approach provides an excellent foundation for modular development based on models, which is the final product of the architect that functions as a set of blue-prints for software developers to follow [Maier and Rechtin, 2000].

Various Development Viewpoints

In this thesis, we are concerned with the software architectural viewpoint via *architecture-driven software development*, although this is one of four viewpoints available for development. In this section, we introduce the development (Table 2.6) pertaining to

Management view using Rational Unified Process (RUP), *Software Engineering view*
with stepwise refinement using waterfall and spiral model, *Product Engineering view* using the three stages to product design of Pahl and Beitz [Pahl and Beitz, 1988], and *Software architectural view* of Sewell called Architecture-driven Software Architecture
[Sewell and Sewell, 2002; Albin, 2003; Booch, 1999; Jacobson et al., 1999; Kruchten,
1999].

Table 2.6. Differentiating development viewpoints.

Viewpoint	Characterization
Management Develop-ment	Development characterized by emphasis on man-agement addressing software development process issues and product life cycle needs, e.g. Rational Unified Process (RUP): best practices for software development teams, a Software Engineering Process that provides a disciplined approach to assigning tasks and responsibilities within a development or-ganization.
Software Engineering Development	Development characterized by emphasis on re-quirements analysis, Design, Implementation, De-ployment, Maintenance
Engineering Design De-velopment	Development characterized by emphasis on phases of Conceptual, Embodiment, Detail design
Architecture Development	Development characterized by emphasis on *archi-tecture-driven software construction* in phases: *Pre-design phase, Domain analysis phase, Schematic design phase, Design development phase, and Build phase.*

Phases of Architecture-driven Software Construction

In *architecture-driven software construction,* the architectural view of a software
development cycle concentrates on domain-specific application development and the architectural development [Sewell and Sewell 2002]. The architectural development (Table 2.7) is sequential, milestone-driven, and divided into four phases: *Predesign phase,*

Domain analysis phase, Schematic design phase, Design development phase, and Build

phase [Albin, 2003].

Table 2.7. Architecture-driven software construction.

Phases	Description
Predesign phase	Formulation of application context
Domain analysis phase	Application requirements are analyzed and structured
Schematic design phase	Demarcation of system structure and modules, dependencies, and rationale
Design development phase	Refinement and variations generated for domain-specific suitability
Build phase	Software engineering and documentation usually with hired contractors

STANDARD ARCHITECTURAL DESCRIPTION TERMINOLOGY

The product of an architect is ultimately a set of blue-prints, or models, depicting various views and viewpoints [Maier and Rechtin, 2000]. We will discuss models and views, the various methods of developing them, and architecture development views from the four main viewpoints of architecting: management, software, engineering design, and architecture perspective [Albin, 2003].

Generic Models, Views, and Viewpoints

ANSI/IEEE Std. 1471 provides a recommended description for the state-of-the-art terminology of architecture [IEEE-1471]. There are variations in terminology, and terms that are important for developing a modeling framework: model, view, and viewpoint (Table 2.8).

Table 2.8. Basic architecture terminology based on IEEE 1471.

Terminology	Description	Reference
Model	approximation, representation, or idealization of selected aspects of the structure, behavior, operation, or other characteristics of a real-world process	IEEE 610.12-1990
View	a representation of a system from the perspective of related concerns or issues	IEEE 1471-2000
Viewpoint	template, pattern, or specification for constructing a view	IEEE 1471-2000

Architecture Development Models

Architecting is an iterative, as well as recursive process that needs *step-wise* development. Stepwise model refinement is a software development approach based on a step-wise, or progressive, removal of abstraction in models, evaluation criteria, and goals, connecting the conceptual model to the more concrete engineering processes [Maier and Rechtin, 2000]. Similar to step-wise refinement, we define the *waterfall model* [DOD-STD-2167A/498, 1995], *spiral model, and agile software development model* [Boehm, 1988] (Table 2.9).

Table 2.9. Software development models.

Method	Description
Waterfall Model	Software development model going through phases: requirements analysis, design, implementation, validation, integration, and maintenance
Spiral Model	Software development model combining elements of both design and prototyping-in-stages, emphasizing risk mitigation
Agile Model	Software development principled on adaption, based on rapid application development (RAD), replaces the traditional waterfall cycle with a repeating series of *speculate, collaborate,* and *learn* cycles

Software Architecture Categories

In 1995, the *First International Workshop on Architectures for Software Systems* provided Shaw the opportunity to distill and categorize the various architecture types used in academia [Shaw, 1996]. In fact, the foundation of architectural terminology can be traced in part to the team effort at Carnegie Mellon, where Shaw served in the IEEE architectural working group [IEEE 1471, 2000], as well as Chief Scientist at the Software Engineering Institute [SEI, 2006]. According to Shaw, software architecture can be divided into five models defining views of a system, namely (1) *Structural Model* showing the constituent parts (2) *Framework Model* showing the whole entity (3) *Dynamic Model* showing behavior, and (4) *Process Model* showing the approach to construction and integration with other models (5) *Functional Model* showing model functionality (Table 2.10).

Table 2.10. Architecture categories.

#	Model Type	Description
1	*Process Model*	Provides methodology to mesh framework and structural model (e.g. Axiomatic V-Model)
2	*Framework Model*	Provides an overall representation of architecture (e.g. Acclaro DFSS with ADT)
3	*Structural Model*	Provides specific OO views (e.g. Telelogic TAU with UML)
4	*Dynamic Model*	Provides OO views expressing interactive and dynamic parts (e.g. Telelogic TAU/UML)
5	*Functional Model*	Provides views expressing functionality (e.g. Acclaro DFSS with ADT)

Software Architecture Styles

Architecture models have various *styles* (Table 2.11) [Maier and Rechtin, 2000].

Table 2.11 Architecture style definitions.

Architectural Style	Definition
Pipe and filter architecture style	Containing one type of component, the filter, and one type of connector, the pipe
Object-Oriented architecture style	Built from components that encapsulate both data and function and exchange messages
Event-based architecture style	An event-based architecture has its fundamental structure a loop which receives events in the context of a system state, and takes actions based on the combination of the event and state
Layered architecture style	Emphasizes horizontal partitioning of the system with explicit message passing and function calling between layers.
Blackboard architecture style	Built from a set of concurrent components, which interact by reading and writing asynchronously to a common area.

Modular Framework Construction

Architects can follow a set of guidelines when developing a framework with modules (Table 2.12) [Maier and Rechtin, 2000].

Table 2.12. Modular framework construction guidelines.

Architectural module characteristic	Recommended architecting guidelines
Module Fan-in	Should be maximized
Module Fan-out	Should generally not exceed 7(+/-)2
Module Coupling	Should be coupled with respect to data, data structure, control, common, content
Module Cohesion	Should exhibit cohesion with respect to functional/control, sequence, communication, time, periodic, procedure, logic, and coincidence

EMERGENCE OF NEW ARCHITECTURE FRAMEWORKS FOR KBE SYSTEMS

Previous Frameworks did not benefit from Model-driven Architecture Approach

Architectures providing a structure for automatically managing domain-specific product engineering knowledge for design support have been done in the past for various applications through KBE techniques [ICAD Release 7.0, 2004; Manjarres et al., 2002]. This type of architectural framework would only encompass an *instantiation* of a KBE system for a *particular* domain. Usually these type of systems are not adaptable, scalable, and rarely reconfigurable, so they quickly become legacy systems, especially due to technology obsolesce and lack of good knowledge modeling for any given application. Many of their architectures and designs were developed according to the technology of the time and not on model-driven approaches, which resulted in systems that quickly became outdated due to current technology emphasis.

Rapidly Advancing Technology Outpacing Legacy Systems

When no consideration is made to advancing technology, the benefits of late-binding of software to architecture models cannot be fully realized. In fact, software architecture itself was an emerging discipline in 1995 [Shaw, 1996] and Model-driven architecture development providing technology independence was not even on the horizon until 2001 [OMG, 2006]. As a result, many artificial intelligence applications have been perceived as failures in the past, partly because the applications did not have a solid foundational architecture to properly structure their complex mechanisms. Furthermore, previous architectures did attempt to leverage the new technologies available through global knowledge acquisition techniques, delimiting their knowledge base only to local sources. In fact, many of these technologies, such as the Semantic Web, Web Services, Internet2, and Next Generation Internet [NGI, 2006, W3C, 2006] simply did not exist, so they were not even considered in previous architecture development.

OBJECT-ORIENTED MODELING LANGUAGE EXTENDING FRAMEWORK

Architectural Modeling Language for Object-Oriented Specification

Systems Modeling Language (SysML) expands Unified Modeling Language (UML) notation, the defacto modeling language for software engineering [INCOSE, 2006b; Booch et al., 1999]. SysML is a domain-specific, object-oriented architectural modeling language specification developed for systems engineering, extending Unified Modeling Language (UML) (Fig 2.5). This extension enables software engineering and systems engineering expressions to merge for common development. The modeling language supports the specification, analysis, design, as well as automated verification and validation of a broad range of systems and systems-of-systems. SysML is supported by many in industry and academia, including defense contractors such as Lockheed Martin, Northrop Grumman, and Raytheon.

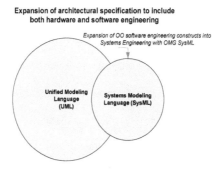

Fig 2.5. SysML extends UML 2.0.

SysML Recent Adoption by Object Management Group

A strong proponent of this language, the International Council for Systems Engineering (INCOSE) is heavily involved in its development and promotion, working with industry to widen its impact. The most current version of SysML was recently minted by the OMG, with a vote to request adoption of the specification was unanimously at the St Louis meeting as recent as April 26, 2006 [INCOSE, 2006a]. Although far from perfect, the language has excited the modeling community by opening the floodgates for software engineers to communicate across-the-board with those engineers working in hardware. The ramification of this include the prospect of development of architecture at the object-oriented level with both software and hardware components using a commonly understood, widely accepted global engineering language.

SysML Provides Blueprints for Both Software and Product Design

SysML is a visual modeling language endowed with a more defined structure for architectural description for an engineering product having both software and hardware components. In addition to the standard UML constructs [Gogolla, M., 1998] such as behavior (defined by the activity diagram), sequence diagram, state machine diagram, and use case diagram, and the structure diagram (defined by the block definition diagram), internal block diagram, and package diagram, SysML has been modified with a requirements diagram and parametric diagram. During the design process, the product architecture can be determined (and later automated with KBE), followed by further refinement of the objects that the architecture structures for the particular product, as in optical backplane engineering or spacecraft design. In fact, a methodology is presented that can be automated for KBE applications for developing product architecture, where the component interactions are actually considered after the architecture is selected [Ulrich and Eppinger, 2004]. Hence, SysML provides an excellent foundation for expression of the final blueprints for a design in the form of OO specifications. These blueprints can be further enhanced (and readily automated with KBE techniques explained later) with clusters of components within the primary structural foundation provided by the SysML language (Fig. 2.6).

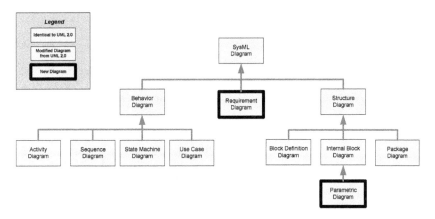

Fig. 2.6. SysML Language structure.

ARCHITECTURE RISK MITIGATION AND RELIABILITY ENGINEERING

SYSTEMS LEVEL RISK MITIGATION PROCESS

Defining Risk

Modeling and risk identification are very important considerations during the design process. The Office of Safety and Mission Assurance at NASA defines risk as "a feasible detrimental outcome of an activity or action…subject to hazards" [NASA, 2000b]. The Systems Engineering Handbook defines risk as "A measure of the uncertainty of attaining a goal, objective, or requirement pertaining to technical performance, cost, and schedule", where the level of risk actually depends on two related factors of likelihood of occurrence and consequences [INCOSE, 2006a]. *Risk* has widely been quantified as a measure of the probability and severity of adverse effects, a subjective metric which has not changed much in the past four decades [Lowrance, 1976].

Risk Mitigation Impact Scope

　　Academia, government, as well as industry, are impacted by risk. In addition to risk standards developed by INCOSE for industry, NASA has implemented Probability Assessment Techniques (PRA) throughout all its programs using techniques such as Fault Tree Analysis (FTA) and Failure Mode and Effects Analysis (FMEA), as well as Design Structure Matrix (DSM). In academia, the Center for Risk Management of Engineering Systems (CRMES) at the University of Virginia, led by Lambert, has done much work in the area of risk mitigation impacting many areas: vulnerability assessment based on entropy, large-scale complex hierarchical systems, critical infrastructure interdependency analysis, safety-critical systems, reliability modeling of multiple failure modes in complex systems [Lambert et al., 2005; CRMES, 2006]. Since many complex systems are vulnerable to multiple sources of risk, methodologies have been developed to assess and quantify level of vulnerability between incident reports and large-scale systems [Lambert et al., 2005]. The assessment of risk has been explored with entropy. Shannon introduces information entropy as a measure of the information content conveyed by a message [Shannon, 1948; Sloane and Wyner, 1993; Chatzigeorgiou and Stephanides, 2003]. A framework that provides a means to use entropy as a measure of risk was also introduced later [Valishevsky and Borisov, 2003].

Risk Management Steps

At NASA Johnson Space Center, Perera summarizes the goal of risk management is to "identify what can go wrong, how likely it is for these to occur, and what are the consequences if they were to occur" [Perera, 2002]. A risk assessment and management process can be developed in the following five steps: risk identification, risk quantification and measurement, risk evaluation, risk acceptance and avoidance, and risk management [Haimes, 1981]. Similarly, the updated 2004 version of the Risk Management Process is outlined by the Systems Engineering Handbook is Planning, Risk Identification, Risk Assessment, Risk Analysis, and Risk Handling [INCOSE, 2006a] (Table 2.13).

Table 2.13. Systems engineering risk steps.

Risk Step	Description
Planning	Establishment of the risk management plan and assignment of responsibilities
Risk Identification	Anticipation, recognition, prioritization of potential adverse outcomes and the associated root causes
Risk Assessment	Characterization of the magnitude and likelihood of risks
Risk Analysis	Evaluation of costs and benefits associated with available risk mitigation methods
Risk Handling	Intervention for elimination and reduction of risk and tracking for assurance and verification

Calculating for Risk Assessment

The NASA, INCOSE, and the UK Ministry of Defence all have a similar method for assessing risk by taking into account two factors: event impact and likelihood. Furthermore, based on the INCOSE Systems Engineering Handbook definition of risk, a subjective rating of risk is proposed called *Expected Value Model*, where risk is expressed as: *Expected consequence = Probability of failure (Pf)* Consequences of failure (Cf)* [INCOSE, 2006a]. NASA [PRA, 2006] definition of risk is very similar to the definition provided by that of INCOSE and the United Kingdom, Ministry of Defence, which defines risk "as the scale of change propagation predicted between subsystems is measured as probabilistic cost, or risk, which is defined as the product of the *likelihood* of the change occurring and the *impact*, or cost of subsequent change" [Defstan, 2006] (Fig. 2.7). As the likelihood of failure increases, risk increases; as the impact of failure increases the risk increases. However, if both the likelihood and impact increases simultaneously, then the risk becomes compounded, which is reflected by a multiplier affect.

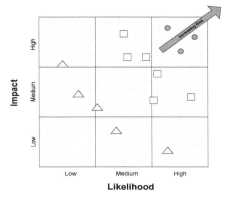

Fig. 2.7. Risk assessment calculated as product of likelihood and impact.

Probabilistic Risk Assessment applied by NASA

In evaluation of risk, "NASA is guided by the Probabilistic Risk Assessment Procedures Guide for NASA Managers and Practitioners" [NASA, 2002]. Many of these procedures can be automated. In the case of the International Space Station (ISS), a report at the Johnson Space Center has identified Probabilistic Risk Assessment as a critical approach to risk management. The report elaborates on PRA as a "comprehensive, structured and logical analysis method for identifying risks in complex technological systems for the purpose of cost-effectively improving their safety and performance...the PRA results can be used as a powerful decision-making tool in support of design, operations, and prioritizing upgrade or reconfiguration decisions. The process helps identify potential new risks, analyze existing risks and subsequently can weigh different options available to the Program to mitigate those Risks..." [Perera, 2002]. In any risk assessment project such as the Cloudsat project, a combination of FMEA and FTA, types of PRA technique, can be applied and automated with KBE techniques [Basilio et al., 2000].

SYSTEMS ENGINEERING APPLIED TO PRODUCT DESIGN RELIABILITY

Defining Reliability

Reliability of a system can be defined as the probability that the system will perform its required function throughout a specified time period [Ertas and Jones, 1996]. One aspect of risk management that pervades safety-critical engineering, as experienced by NASA for instance, is the emphasis on reliability, which also relates to fault detection, and vulnerability assessment, and risk mitigation techniques under the term Probability Risk Assessment [NASA, 2000a]. Although such risk mitigation methods have been researched and employed by government agencies, as well as industry and academia, to assess system reliability and vulnerability for safety-critical systems, much work is still needed to intelligently automate them for effective risk management [Trevino et al., 2005].

Systems engineering for reliability

Systems engineering provides many tools and methods for risk mitigation to support the design process [Lewis, 1987]. In this section, we will show the importance of reliability engineering (Table 2.14), the application of systems engineering to the design process for reliable product development. Introducing systems engineering for risk mitigation provides a methodic basis for introducing intelligent automation techniques later. The foundation for orchestrating all of these complex systems is provided by a KBE system.

Table 2.14. Reliability engineering is one of many systems engineering applications.

System Engineering Type	Definition
Reliability engineering	*Assures system will meet expectations for lifetime performance*
Human Factors Engineering	Integrates human element as an explicit part of an ergonomic system.
Control systems design	Involves the development of methods for analysis of control process
Interface design	Ensures interface design and specification are interoperable
Operations research	Investigates optimization under multiple constraints
Safety engineering	Identify safety hazards in emerging designs risk mitigation
Security engineering	Integrates control systems design, authentication, reliability, safety
Software engineering	Involved in designing, creating, and maintaining software
Supportability Engineering	Concentrates on designing a program for system maintainability

STRUCTURAL INTEGRITY OF COMPONENTS FOR RELIABILITY ENGINEERING

Reducing Design Complexity for Increased Reliability

In product development, systematic design risk management applied to reliability engineering is a foundation for intelligent automation of decision support for reliable product design. Simon defines hierarchical decomposition and the complexity of a product in terms of the connections between its parts [Simon, 1981], whose individual component and clustered system reliability is critical for total system function. In order achieve design risk mitigation, product engineering needs to take into account four layers of design complexity [Earl et al., 2004]. The process of reliability engineering for design risk mitigation addresses these concerns for product engineering in a systematic way that can be automated (Table 2.15).

Table 2.15. Layers of design complexity.

Layer	Description
Layer 1	Product design has many components that are highly interrelated and linked in various ways
Layer 2	Process of product design has many risky interlinked tasks with high propagating impact
Layer 3	Organization for product design consists of many multidisciplinary development teams
Layer 4	Product design has complex inter-relationship with its environment

Critical Design Parameters for Reliable Product Development

Although systems engineering can be applied to many other aspects of engineering, our primary focus in this thesis is on improving reliability engineering (Table 2.16). This type of systems engineering approach to risk management applied to product design is *reliability engineering*, which is closely related to system dependability, maintainability, availability, optimization, and cost [Ertas and Jones, 1996]. Hence, our objective is to improve the design process of product engineering for achieving reliable systems, which can be achieved through design risk mitigation techniques applied throughout the design process (Fig 2.8).

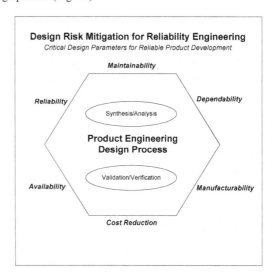

Fig. 2.8. Design risk mitigation for reliability engineering.

Table 2.16. General reliability terms defined in relation to reliability engineering.

Reliability terms	Description
Reliability	Probability that a component will successfully perform in varying conditions
Availability	Probability that the component will be available for operation in given time period
Cost Reduction	Economic impact of component by optimizing for cost
Manufacturability	Degree to which a component is readily manufactured
Dependability	Measures component condition in terms of reliability and maintainability

COMPLEX SYSTEMS

Defining a Complex System

There have been many ways of defining *complex systems* over the past decade. Corning claims that one broadly accepted definition for measuring system complexity does not exist, primarily due to the fact that evaluation of complexity is subjective [Corning, 1995]. Furthermore, the lack of a consensus on a definition for complex systems has been attributed to the area's dependence on perception based on observer judgment, and background [Bearden, 2000]. We provide a table summarizing various definitions for complexity in different fields (Table 2.17).

Table 2.17. Definitions of complex systems.

Field	Emphasis of Definition	Reference
Defense	The ability to network a very large number of entities and very diverse entities	[Cebrowski, 2004]
Axiomatic Design Theory	Amount of Coupling and Information content	[Suh, 2001]
Natural Sciences	The amount of information necessary to describe the system	[Bar-Yam, 2003]
Systems Theory	A multidimensional, multidisciplinary concept	[Corning, 1995]
Systems Engineering	The number and kinds of relations between components	[Hall, 1963]
Chemical Engineering	Large number of components which may act according to rules that may change over time and that may not be well understood; the connectivity of the components may be quite plastic and role may be fluid	[Amaral, 2004]

Complex System Features and Evaluation Dimensions

We identify and define the features of complex systems (Table 2.18). We have integrated and expanded the complexity features tables in [Mahafza et al., 2005].

Table 2.18. Some features and evaluation dimensions of complex systems.

Feature	Description	Evaluation Dimensions
Cause/Effect	Cause and effect are not obvious and direct.	Degree of separation between causes –effects
Design Decisions	The implications of design decisions are not bounded and cannot be easily ascertained or predictable	Number of alternative solutions
Elements	The number and variation of elements is large.	The number and diversity of components
Energy	The resources needed to develop each element are large.	The amount of energy expended
Interactions	The diversity and number of interacting hierarchical elements is large	The diversity of dynamically interacting components
Risks	Risks have unforeseen multiple factors and emergent properties.	Level of risks dominated by system risks
Technologies	Requires mix of different technologies and high coupling between several different technologies.	The diversity of required technology choices and index derivations
Maintenance	Requires a set of diverse standards, techniques, and technologies to maintain system	The diversity of required standards and techniques
Environment	Requires a varied set of environments to function	The diversity of environments addressed

COMPLEX SYTEMS IMPACTING STRUCTURAL RELIABILITY

Maintaining Structural Integrity for Complex Engineered Systems

The *structural integrity* for complex systems can be achieved by emphasizing system reliability and risk mitigation during the design process. This field of reliability is addressed in systems engineering, where individual components, as well as their structures and interfaces are examined to minimize risk, especially during the integration process where most errors occur. Since a failure of one component can lead to system-wide catastrophic failure due to component dependencies, reliability tools and techniques have been developed to address both component and cluster vulnerabilities [Trewn and Yang, 2000]. The increased functionality of modern engineered systems often results in the increased complexity, which would most likely decrease structural reliability. Research has confirmed an inverse correlation between functionality and reliability in complex systems [Bar-Yam, 2003; Schneidewind, 2002] (Fig. 2.9).

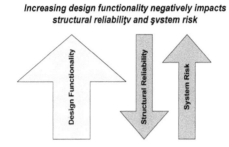

Fig. 2.9. Higher design complexity means less reliability with increasing risk.

NASA RISK MITIGATION APPROACH FOR COMPLEX SYSTEMS

NASA Motivation for Multidisciplinary Research on Complex System Failure

By experience, it has been shown that complex systems also have higher failure rates, resulting in unnecessary delay or even abandonment of expensive projects [Bar-Yam, 2003; Bearden, 2000]. For instance, an analysis conducted on 40 NASA small-spacecraft missions determined the average "complexity" for the failed missions was higher than that of the successful missions [Bearden, 2000], confirming the need for further research into risk mitigation for complex systems. In the case of the Space Shuttle Orbiter, the system modules having excessive size and design complexity has resulted in propagating faults and failures [Schneidewind, 2002]. These mishaps provide a basis to study complex systems as a multidisciplinary effort with some relatively new methods being proposed [Calvano and John, 2004]. In addition to these failures that have taught important lessons in the space program, two unexpected Orbiter tragedies were experienced that spurred even more emphasis on risk mitigation. Despite attempts, there is still no set of general criteria or metrics that can be applied to engineered systems for complex systems [Mahafza et al., 2005]. However, one way to attack the problem of complex systems is through risk mitigation techniques that can be automated through KBE methods, which we will discuss later.

General Reliability Measures Employed by NASA

Government, industry, and academia have been developing and sharing reliability measures and techniques for risk management with NASA (Table 2.19). The space program has a history of references showing the pervasiveness of risk mitigation techniques based on Probability Risk Assessment (PRA), such as Failure Mode and Effects Analysis (FMEA) and Fault Tree Analysis (FTA), which can be found in the government space program report NASA/SP—2000–6112 [NASA, 2000a]. Since design complexity is an important consideration that could impact system reliability, preferred reliability practices based on Reliability Block Diagrams (RBD) have been certified by NASA Johnson Space center in PRACTICE NO. PD-AP-1313 for the Orbiter and Space Station missions [NASA JSC, 2005]. NASA's Software Assurance Technology Center (SATC) located at Goddard Space Flight Center (GSFC), has introduced metrics for object-oriented code evaluation in terms of classes, methods, cohesion, and coupling [NASA GSFC, 1998].

NASA has shared many techniques for reliability and quality assurance with industry and academia for both product and software engineering. For instance, Six Sigma
[Wortman, 2001] companies also utilize techniques such as FMEA and Quality Function
Deployment (QFD) [Wortman, 2001]. In academia, Axiomatic design theory [Suh, 2001]
has been endorsed by the National Academy of Engineering in 2001 and National R e-
search Council [NRC, 2006] and related software, namely Acclaro DFSS, has been acquired by aeronautical engineering companies, such as General Dynamics [Axiomatic
Design, 2005]. Design Structure Matrix (DSM) has been shown to be an excellent approach for analyzing structural interfaces and resolving component dependences
[Pimmler and Eppinger, 1994]. For instance, in his MS Thesis at Massachusetts Institute
of Technology (MIT), Brady applied DSM to analyze a multitude of NASA missions in
terms of component risk factors impacting structural dependencies and interfaces in *Utilization of Dependency Structure Matrix Analysis to Assess Implementation of NASA's
Complex Technical Projects* [Brady, 2002].

Table 2.19. Reliability analysis methods for complex systems management.

Reliability Analysis Method	Application
Fault Tree Analysis (FTA)	Identification of system faults and vulnerabilities
Failure mode and effects analysis (FMEA)	Identification of potential risks and preventive methods
Probability Risk Assessment (PRA)	Methodology traditionally employed by NASA which includes FTA, FMEA, etc
Design structure matrix (DSM)	Component dependency and risk area resolution
Quality function deployment (QFD)	Assurance that qualitative needs are addressed by functional requirements
Axiomatic Design Theory (ADT)	Reducing complexity by decoupling dependences and identification of alternative designs based on information content
Theory of Inventive Problem Solving (TRIZ)	Resolving design trade-offs by innovation
Reliability Block Diagram	Calculating interconnected component risk
Object-Oriented Code Evaluation	metrics for object-oriented code evaluation in terms of classes, methods, cohesion, and coupling

Current NASA Intelligent Automation Techniques for Reliability

An increased level of protection against catastrophic failure could be made provided by introducing advanced risk mitigation techniques based on KBE techniques.
The importance of many of these reliability measures introduced in this thesis is due to
their unique suitability for automation via artificial intelligence techniques. At NASA
Marshall Space Flight Center (MSFC), we were introduced to state-of-the-art application
areas for reliability automation by the artificial intelligence group working on automated
rocket health assessment [Trevino et al., 2005]. The architectural framework in this thesis
and many of its modules were developed based on current and projected NASA needs,
within the scope of an academic thesis providing a platform for configuring KBE systems
to support the design process for product engineering – *with particular emphasis on risk
mitigation*. Thus, we were able to expand the application of the thesis case study on optical backplane engineering to include NASA requirements as well. In this thesis, in addition to the architectural framework for developing a class of KBE systems for design
support, we show how this type of reliability analysis identifying vulnerabilities and potential risk areas can be reformatted into data for machine processing and intelligent
automation through *knowledge engineering* techniques for eliciting and capturing domain
knowledge. More on this topic will be discussed in the artificial intelligence survey and
other chapters after the review on reliability. The AI group at NASA Marshall Space
Flight Center has considered implementing many different versions of the intelligent
automation techniques based on KBE techniques as computing power has advanced
[Trevino et al., 2005]. Many of these techniques have been shared across industry and
academia, as well as government.

SYSTEMATIC DESIGN RISK MITIGATION AUTOMATION METHODS

SYSTEMATIC DECOMPOSITION AND MAPPING OF DOMAIN REQUIREMENTS

Domain-Based Requirements Analysis in Axiomatic Design

Generally speaking, the most functional complex systems in nature exhibit two main properties (1) hierarchical structure and (2) dependency minimization [Maciaszek, 2004]. In parallel, Axiomatic Design Theory, developed at MIT, also provides a methodology for hierarchical decomposition and dependency resolution by application of axioms, corollaries, and theorems to decouple design and reduce their complexity be minimizing their dependencies [Suh, 2001]. All of these axiomatic applications are accomplished in four separate domains (Fig. 2.10): *Customer, Functional, Physical, and Process,* each of which is defined by a set of customer attributes (CA), Functional Requirements (FR) Design Parameters (DP), and Process Variables (PV) [Suh, 2001].

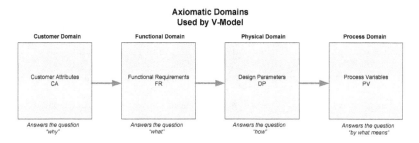

Fig. 2.10. Axiomatic Domains used by V-Model.

Hierarchical Decomposition of Domains

Hierarchical decomposition is present in all complex systems in nature, economics, as well as engineering systems [Simon, 1981]. Based on customer needs, an architectural description can be hierarchically decomposed into hundreds of individual functional requirements, each of which can be mapped to a set of design parameters to form a design matrix (Fig. 2.11). Once formulated, this matrix can be used to decouple elements for dependency resolution using methods such as DSM [Albin, 2003]. The V-Model provides a method for mapping this high level architectural framework model development to object-oriented structural models based on standard UML diagrams as a precursor to software development using this architecture approach [Suh, 2001].

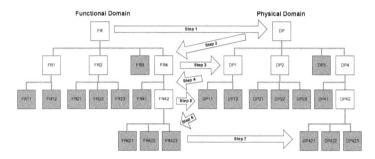

Fig. 2.11. Hierarchical Decomposition of Axiomatic FR and DP.

Axiomatic Approach to Decomposition Provides Inherent Verification

Modern axiomatic design approach has supplanted a software design methodology popular in the mid-1970s and 1980s which used similar hierarchical decomposition with a top-down approach, such as Structured Analysis and Design Technique (SADT) [Ross, 1985]. However, SADT decomposition analysis is done only in the physical domain, without a zigzagging process that maps system functional requirements to design parameter accomplished in axiomatic design in both the functional and physical domain, assisting in verification. Thus, Suh's axiomatic approach takes into account the customer needs, important to requirements traceability, at every step of functional requirements breakdown during the process of decomposition.

47

SYSTEMATIC INTERACTION ANALYSIS FOR DECOMPOSED SYSTEM

Applying Design Structure Matrix to Decomposed System

Once a system is hierarchically decomposed, systematic integration of these parts can be achieved by streamlining for modularity using a design structure matrix (DSM) that represents component interactions in several ways, such as information, spatial, material, and energy exchange [Pimmler and Eppinger, 1994]. For representing dense graphs, the Design Structure Matrix (DSM) offers a more compact form for a more manageable display [Ghoniem et al., 2004]. The diagram represents any given set of components that interact with each other (Fig. 2.12), which can be represented in matrix format (Table 2.20). Eppinger shows how to decompose the system, document the interactions between elements, and cluster them into chunks, which define the product architecture and system team structure assigned to developing the product. A technology risk factor also has been introduced as a multiplier that takes into account the technology readiness level of each component based on NASA criteria in an MIT MS thesis applying DSM to NASA missions [Brady, 2002].

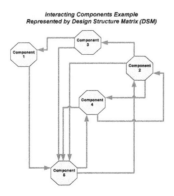

Fig. 2.12. Interacting component diagram that can be represented by DSM.

48

Table 2.20. Component DSM example with five components.

	C1	C2	C3	C4	C5
C1	X				X
C2		X	X	X	X
C3	X		X		X
C4		X		X	X
C5		X		X	X

SYSTEMATIC FAULT ANALYSIS FOR PROBABILITY RISK ASSESSMENT

Fault Tree Analysis as a Basis to Conduct Probability Risk Assessment

Although fault tree analysis is not a new technique, it has widespread application in NASA and industry today. As early as 1964, its relevance to analysis of potential faults in high-risk projects is apparent with its application to Minuteman Inter-Continental Ballistic Missiles (ICBM) developed by Boeing Corporation [NASA, 2000a]. This type of probabilistic risk assessment technique can be adapted to analyzing potential faults occurring during the design process, especially since it is useful for functional analysis of complex systems. Evaluation of system reliability can be accomplished with FTA by tracing potential design defects to a particular component. The Nuclear Regulatory Commission Fault Tree Handbook (Table 2.21) defines fault categories along four distinct categories [NRC, 2006]. As can be seen, there are four levels of faults, ranging from a Level I, a zero impact fault, to Level IV, a catastrophic impact fault. These types of fault standards already established for safety-critical systems can be used as a basis to develop automated systems in the future that have correlated inference weighting.

Table 2.21. Nuclear regulatory commission fault tree handbook definitions.

Level	Fault Description
Level I	A negligible fault producing no impact to system performance or capability/capacity
Level 11	A fault reducing overall system capacity, but does not impact on-line performance
Level 111	A fault reducing overall system capacity and degrading on-line performance
Level IV	A catastrophic fault rendering the system being analyzed as completely non-operational.

Prospect of FTA Technique Automation for Tracing Faults to Source

A system-level fault tree can be constructed using AND/OR gates to identify fault scenarios that could cause structural failure during the design process [Basilio et al., 2000] (Fig. 2.13). This approach is very conducive to automation using KBE techniques. The automatic tracing of the fault and the failure probabilities can be enhanced with FMEA, and can be automatically calculated using KBE techniques during the design process. Domain-specific component structures for each fault type can be determined per case study, such as for the engineering of an optical backplane, as in the case of the Mars Pathfinder unmanned probe [Brady, 2002].

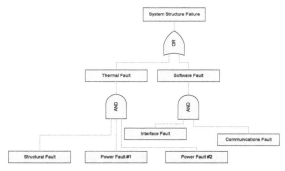

Fig. 2.13. FTA Technique for tracing fault to source.

Prospect of FTA Technique Automation for Calculating Failure Probability

A risk rating can be constructed that can be applied to the FTA paradigm to calculate the probability of each fault occurring to assess the overall impact on the system (Table 2.22). This type of calculation is ideal for automation using KBE systems due to its need in any engineering design process, in addition to its highly quantitative determination method and reliance on large knowledge repositories. When a component probability level reaches 1-E3 and I-E4, this considered to be the threshold for many active mechanical parts (such as actuators and springs), and the threshold for active electrical parts (such as relays and transistors). Many of these probabilities can be stored in the knowledge base and adjusted as needed. Calculation is done by multiplication of probabilities at the AND gates and addition at the OR gates for total failure probability risk assessment using FTA. This risk rating was obtained from work at NASA JPL based on the Cloudsat project [Basilio et al., 2000]. As can be seen (Table 2.22), one can see that the fault probability becomes increasingly less by a factor of 10, at every increment of level. This type of data is ideal for populating a searchable database with critical factors affecting the likelihood of success for each mission.

Table 2.22. Risk rating applied to each component for structural risk calculation.

Level	Rating	Fault Probability
0	1E-0	1 in 1
1	1E-1	1 in 10
2	1E-2	1in 100
3	1E-3	1 in 1000
4	1E-4	1 in 10,000
5	1E-5	1 in 100,000

SYSTEMATIC INNOVATION FOR RELIABILITY ENGINEERING

Systematic Innovation Introduced by the Theory of Inventive Problem Solving

An engineered complex system can be represented as a hierarchically decomposed structure [Simon, 1981] which consists of many major interacting subsystems composed of functional components [Trewn and Yang, 2000]. During all phases of engineering, critical decisions need to be made to resolve design trade-offs that can jeopardize system reliability. One popular industrial approach to resolving such trade-offs methodically is by the systematic innovation process proposed by Altshuller detailed in the Theory of Inventive Problem Solving (TRIZ) [Suh, 2001]. This approach can be used to resolve trade-offs by identifying innovative solutions already made in other fields and adapting them to the present design in progress by eliminating the design trade-off altogether [Altshuller, 1997].

Combining FTA and TRIZ for Design Trade-Off Analysis to Resolve System Faults

The TRIZ approach [TRIZ, 2006] is ideal for increasing overall reliability of a design by providing a systematic means to introduce a series of innovations throughout the conceptual, embodiment, and design phases. Complex systems usually have more faults and are more difficult to maintain [Schneidewind, 2002], requiring a tremendous amount of analysis to prevent propagating faults. There are many types of analysis methods for analyzing these faults, such as Fault Tree Analysis (FTA), Failure Mode and Effects Analysis, Design Structure Matrix, and TRIZ (DSM) as provided by probabilistic risk assessment methods used by NASA. Of these methods, TRIZ is the most useful for introducing systemic innovation into analysis, which can prevent design faults from occurring in the first place. This analysis is done by identifying defects and the associated design decisions that were made, followed by resolving the design trade-offs that were made early in the design process that could have prevented these defects from occurring in the engineered product. Using the other PRA methods, such as FTA, the fault can be traced to its source and TRIZ can be applied to identify an innovative design parameter to replace the previous one causing conflicting dependencies resulting in failure or propagating fault, for instance.

TRIZ Obtains Best of Both Worlds by Elimination of Design Trade-Off

By leveraging TRIZ techniques, similar conceptual solutions can be identified in totally unrelated fields in the patent database and be applied to resolve a particular design trade-off encountered during the design process. This means that TRIZ trade-off analysis provides a system means for an innovative solution to be discovered from patent searches to resolve the trade-off by simply *eliminating* it, thereby getting best of both worlds. For instance, if during the design process, a decision must be made on design features that are usually inversely related (e.g. speed and weight for aircraft design), TRIZ can resolve it and improve the design by identifying a concept already employed in another field that allows *both* high speed *and* light weight. This trade-off analysis can be accomplished by applying Altshuller's TRIZ methodology that enables feature to feature (e.g. Speed/weight) comparison in a *contradiction matrix* for identification of an innovative solution that can be borrowed from another domain [Altschuller, 1997].

Instance of TRIZ Leveraging Remote Databases and Patent Databases

The resources TRIZ can leverage include patent databases containing millions of pre-existing solutions that are exploitable through this method. TRIZ methodology compares a feature to be maintained without sacrificing another feature based on a 39 by 39 pre-defined matrix, called a contradiction matrix. For instance, in the case of speed/weight, the designer can identify patents where a similar trade-off between the inverse functions of speed and weight were resolved in another domain, such as the military, where it is determined that a special type of alloy was already discovered and is used for tanks (e.g. titanium with special properties of strength and light weight). This discovery would give the designer an opportunity to introduce this solution for increased reliability, while maintaining light weight - and speed. Hence, the design tradeoff itself is eliminated by selecting titanium as his material of choice for his design (for aircraft armor), which already is being used in the military (for tank armor). The importance of this method from an AI standpoint is that the sheer volume of preparative work required to make this approach practical can be accomplished with appropriate KBE techniques. Hence, the TRIZ process can be *automated*.

KBE SYSTEM-OF-SYSTEMS APPROACH FOR DESIGN RISK MITIGATION

SYSTEM-OF-SYSTEMS APPROACH FOR DESIGN RISK MITIGATION

System-of-systems is emerging as an important area to research [Madni, 2006; Madni et al. 2001 and 2005]. *System-of-Systems* (SoS) are essentially large-scale concurrent and distributed systems comprised of constituent complex systems [Kotov, 1997]. The study of System-of-Systems generally encompasses a wide spectrum of interrelated multidisciplinary concepts that exhibit complex systems behavior, including architecture development, design, systems engineering, agent-based modeling, and object-oriented simulation [Sage and Palmer, 1990; Sage and Cuppan, 2001]. We will provide definitions and features of *complex systems* embedded in a SoS paradigm, as well as define the role architecture plays in achieving intellectual control over a SoS KBE application applying artificial intelligence to the design process [Ring and Madni, 2005].

Identifying a System-of-Systems Application

The paradigm of *System-of-Systems* (Table 2.23) is an emerging practice that addresses *large-scale interdisciplinary* problems having multiple, heterogeneous, distributed systems at multiple levels and domains exhibiting *complex* relationships [DeLaurentis, 2005; Boehm, 2006]. Kotov describes SoS as "large-scale concurrent and distributed systems the components of which are complex systems themselves" [Kotov, 1997]. Many large-scale government projects, such as space systems and transportation, usually exhibit a number of SoS traits [DeLaurentis, 2005]. We will define complex systems and their impact on reliability assurance during the design process in later sections.

Table 2.23. Traits identifying systems-of-systems paradigm.

SoS Traits	Constituent systems Description
Operational and Managerial Independence	Constituent systems are useful in their own right and generally operate independent of other systems with unique management intent
Geographic Distribution	Constituent systems are not physically co-located, but communicate
Evolutionary Behavior	Constituent systems sustain continuously changing system boundaries
Emergent Behavior	Constituent systems responsible for unpredictable new behavior
Networks	Constituent systems are comprised of networks defining the connectivity between independent systems via interaction rules
Heterogeneity	Constituent systems are of significantly different nature, with different elementary dynamics that operate on different time scales
Trans-domain	Constituent systems require unifying knowledge across multidisciplinary fields of study

System-of-Systems Approach Ideal for Large-Scale, Complex Systems

Many large-scale engineering projects need a system-of-systems approach for risk management. The risk can be attributed to their inherent complexity, generally characterized by a set of complex systems composed of a high-degree of interdependent, dynamic parts acting together [Stermon, 2000]. As a result, this event motivated the passage of the Information Technology Management Reform Act (ITMRA) in 1996, a part of the Clinger-Cohen Act [Clinger-Cohen Act, 1999]. Many large-scale projects cost between $50 million to $5 billion each and span several years of development [Bar-Yam, 2003]. A general survey of large software engineering projects classified projects by considering if they met critical project objectives, based on a reasonable timetable with cost estimates [The Standish Group International, 1994]. Their results indicated that only 1 in 5 projects actually met all three basic engineering criteria. In fact, when considering the direct costs associated with large-scale projects of this nature, the amount of loss can exceed $100 Billion [Bar-Yam, 2003]. The underlying reason for the difficulty involved in large engineering projects can be traced to a systemic problem of managing complex systems within a system.

SCOPE OF ARTIFICIAL INTELLIGENCE APPLIED TO DESIGN PROCESS

Class of Knowledge-Based Engineering Systems Supporting Design Process

Artificial intelligence is a broad field, which includes expert systems, knowledge-based systems, robotics, and Knowledge-Based Engineering (KBE) systems [Russel and Norvig, 2003]. Our scope of focus for artificial intelligence is limited only to KBE systems applied to the design process and various algorithmic methods adapted for design support and intelligent agent interaction with Web Services. In order to accomplish these dynamic applications, we are developing an architectural framework for a KBE system that manages a set of complex, distributed systems, which we will describe in the next chapter.

Typical Functions of a General Intelligent System Not Specialized for Design

Typical functions of an intelligent system, not specialized for design, are [Cassidy, 2004; Sage and Palmer, 1990] (Table 2.24). Although many of these functions are defined in the table, they are not present in all intelligent systems, as there are many different types of artificial intelligence systems specialized for different tasks. In this thesis, for instance, we focus on KBE systems that support the design process.

Table 2.24. Basic elements of a general intelligent system not specialized for design.

AI element	Description
User Interface	Provides interactive, ergonomic mechanism for user input and feedback
Database	Provides mechanism to record intermediate results during session
Knowledge base	Provides facts plus heuristic planning and problem-solving rules for long-term storage
Inference engine	Provides a mechanism to apply inductive and deductive reasoning
Interpreter	Provides a method to apply the rules
Scheduler	Provides a method to control the order of rule processing
Consistency enforcer	Provides a method to adjust previous conclusions when new data or knowledge is provided
Justifier	Provides a method to rationalize conclusions and explain the system behavior

Comparing AI, Expert, KBE, and CAD Systems

Artificial intelligence is a broad field that is comprised of areas ranging from expert systems, also known as KB systems, to robotics. Our focus is on KBE systems, KB systems which utilize engineering knowledge provided by domain experts for improving the design process. In order for engineering information to be applied, the knowledge based on unprocessed information has to be organized and modeled [Schreiber et al., 1999]. This knowledge can reside everywhere, e.g. on the Internet, in company hardware, or in best practices documents, in global standards, or in humans, usually domain experts. In order for knowledge to support the design process, it needs to be elicited, captured, stored, and represented in a form that a machine can process and manipulate to draw conclusions, through a process of *knowledge engineering.* A field that overlaps with KB systems is *expert systems,* although they do not emphasize the engineering aspect of knowledge-based systems [Norvig, 2003]. A KB system has explicit, declarative description of knowledge for a domain-specific application [Speel, 2001]. In contrast to information systems, there is an increased importance on knowledge modeling [Schreiber et al., 1999]. KBE provides engineering rationale prior to Computer-Aided Design (CAD). Once a design has been generated from a KBE product model, the geometric information can be transferred to a CAD system for further detailing or analysis.

CAPTURING DESIGN RATIONALE FOR RELIABLE PRODUCT ENGINEERING

Criticality of Design Rationale for Justifying Decision-Making

Design Rationale (DR) consists of the decisions made during the design process and the reasons behind them [Ball, et al., 2001]. Capturing and representing DR is an important capability in the design process that can be automated using KBE systems. DR keeps track of design decisions, justifications, and alternatives, and various arguments considered, while assisting in revising, maintaining, documenting, and evaluating the design [Lee, 1997].

The original designer's intent can be analyzed by subsequent designers and domain experts can update the reasoning with improved rational associated with alternative design parameters proposed. Time, money, and resources can be saved by avoiding unnecessary duplication of work that was done in previous design sessions. In fact, capturing design rationale was determined to be very useful during requirements and design, based on a study performed using DR documents to evaluate a design [Karsenty, 1996]. In this study, a full half of the designer's questions gravitated on the actual rationale behind the design. There are many tools and supportive units available for design rationale capture (Table 2.25).

Table 2.25. Sampling of design rationale systems.

DR System	Description	Reference
JANUS	Critiques the design and provide the designers with rationale to support the criticism	[Fischer, et al., 1995]
SYBIL	Conducts conflict mitigation in collaborative design efforts	[Lee, 1990]
Process Technology Transfer Tool (PTTT)	Transfers process design information between development and manufacturing	[Brown and Bansal, 1991]
Device Modeling Environment (DME)	Generates documentation "on demand" about electromechanical devices	[Gruber, 1990]
C-Re-CS	Performs consistency checking on requirements and recommends a resolution strategy for detected exceptions.	[Klein, 1997]

Design Rationale Capture and Representation Forms

Most work on design rationale has concentrated on capture and representation, where the representations range from formal to informal. A formal approach is machine-readable and an informal approach is human-readable. Although manual recording of all decisions made can be time-consuming and expensive, it can be done expediently and cheaply with the right information system application that does it automatically. Design rationale representations range from formal to informal. Design Rationale representations range from informal representations, such as transcripts, audio, or video tapes, to much more formal representations such as rules or ontologies stored and invoked by a KB system [Conklin and Burgess-Yakemovic, 1995]. Three perspectives for design rational can be identified: argumentation, documentation, and communication [Shipman and McCall, 1996] (Table 2.26).

Table 2.26. Design rationale representation forms.

Design Rationale form	Description
Argumentation	Captures the design decisions
Documentation	Captures the reasons for the decisions
Communication	Captures the context in which the design was developed

Capturing Domain Expert Knowledge for Reliable Design Rationale

Oftentimes, a domain expert will not be part of a company forever or is not always available, so a means to capture and demonstrate to colleagues the soundness of his/her decisions and their supporting design rationale is important. This knowledge is recorded for future analysis or modeling, and can serve as a foundation for automation and as basis for developing knowledge models for future KBE applications [Peña-Mora et al., 1995; Pena-Mora and Vadhavkar, 1996]. Furthermore, corporate memory, that is memory developed by an employee over time, could be captured and preserved for future use [Brice and Johns, 1998]. An incentive for this knowledge capture could be due to high employee turnover rate. We present a list of tools to capture design rationale (Table 2.27).

Table 2.27. Useful tools for capturing design rationale.

Tool	Description	Reference
Machine-Learning Apprentice System (M-LAP)	User actions are recorded for application of machine-learning techniques	[Brandish, et al., 1996]
Rationale Construction Framework (RCF)	Uses theory of design metaphors to interpret actions recorded in a CAD tool for conversion into a history of the design process	[Myers, et al., 1999]
Design History Tool (DHT)	Integrated with a design tool, it captures the history as a byproduct of the designing process.	[Chen, et al. 1990]
itBIS and gIBIS	Require that rationale be captured in a specific format	[Conklin and Burgess-Yakemovic, 1995]
Hyper-Object Substrate (HOS)	Uses data captured informally during the design process and converts it into a useable form	[Shipman and McCall, 1996]

Design Rationale Ideal for Rapid Prototyping

Since design is an iterative process where the reasoning for particular decisions may change over time, it is important that the design rationale be updated for iterative design work that may include rapid prototyping. Iterative designs can be analyzed based on previous reasoning and improved iterations can be done continuously. Rapid prototyping can produce design simulations that examine competing prototypes based on different design rationale during design process. This approach provides a way to take into account different paths of reasoning for decisions, leaving room for analysis of various design paths. Sometimes this is critical when one of the branches is determined to be based on invalid or faulty reasoning, which can be recorded as th e most current, Design

Rationale (DR) time-stamped for future analysis. Sometimes, the DR could turn out to be

based more on ergonomic considerations as opposed to actual problem-solving reasons,

which would reveal the underlying bias of the designer. The advantage of shedding light

on the actual thought processes for a variety of design decisions would enable a tighter

and more effective quality control by management using notation (Table 2.28).

Table 2.28. Argumentation notations for capturing design decisions.

Argumentation Notations	Description	Reference
Design Space Analysis (DSA)	Represents via Questions, Options, and Criteria (QOC)	[Maclean, et al., 1991]
Issue Based Information Systems (IBIS)	Represents as issues, positions, and arguments	[Conklin and Burgess-Yakemovic, 1995]
Design Rationale Capture System (DRCS)	Represents using entities and claims about the entities	[Klein, 1993]
Design Recommendation and Intent Model (DRIM)	Represents by capturing recommendation, justification, and intent for each participant in the design process	[Pena-Mora, et al., 1995]
Decision Representation Language (DRL)	Represents InfoRat based SIBYL	[Lee, 1990]

DESIGN PROCESS SUPPORT WITH KNOWLEDGE-BASED ENGINEERING

Knowledge-based Engineering System for Design Process Support

A *Knowledge-Based Engineering (KBE) system* is a KB system that has been

adapted for support of engineering problems, usually by storing and applying pre-defined

design rules and ontologies, whose knowledge base is developed by knowledge engineer-

ing [Schreiber et al., 1999; Stokes, 2001]. Knowledge-Based Engineering (KBE) is the

process of design, testing, and manufacturing of products by systematically capturing and

embedding human domain expertise in a computer and subsequently applying the ma-

chine intelligence to *engineering applications* [Schreiber et al., 1999]. A *domain expert*

is an engineer with years of design experience that is usually in charge of developing and authenticating design rationale over time for the *knowledge base*, which stores the knowledge. Remote *knowledge repositories* are distributed stores of knowledge developed by domain experts anywhere on the globe, and accessed by intelligent agents on demand.

Comprehensive Validation of Knowledge-Based Systems

Validation of KB systems has been a global endeavor. One of the notable efforts for assessing the state-of-the-art of validation for KB systems was accomplished with the three-year VALID Project, conducted by the European Strategic Program on Research in Information Technology (ESPRIT) [ESPRIT, 2000b]. Conducted in the early 1990s, the purpose of the VALID project was to comprehensively assess the emerging field of validation in terms of definitions, techniques, and methods for validating KB systems. After evaluation of this research, it was determined that some of these technologies may be adapted to KBE systems. ESPRIT is run by the Directorate General for Industry of the European Commission. Our comprehensive approach to validation builds on this foundation of research by providing a methodology for validating the Architecture, Design Process, Knowledge Acquisition, as well as Artificial Intelligence aspects for KBE systems, which we will discuss in the validation chapter.

ICAD System: A Functioning Knowledge-Based Engineering System

One of the first and most successful KBE systems developed was the ICAD system, originating from collaboration between Computervision and the MIT Artificial Intelligence Laboratory in the early 1980s [ICAD Release 7.0, 2004]. ICAD was formed to sell the product commercially. While other applications of AI failed, the success of KBE propelled the company into public and became Concentra in 1995. In 1998, the ICAD product line was privatized to form knowledge technologies International, where the ICAD product line was further enhanced. Most design improvement using ICAD has been in the development of KBE applications for designing a family of parts which have impacted the Return On Investment (ROI) in aerospace and automotive design.

KBE Suitability for Engineering Applications

A *KBE system* is a type of intelligent system well-suited for design applications, partly because an Object-Oriented approach may be used for developing the Knowledge Base [Schreiber et al., 1999]. KBE allows for rapid-prototyping of design and systematic ontological development. KBE also provides methodologies for knowledge elicitation and representation through a new approach to knowledge modeling with Moka Modelling Language (MML), an extension of UML [Moka, 2004; Milton, 2004; ESPRIT, 2000a]. A European initiative, ESPRIT Framework IV, developed MML as a framework for structuring and representing engineering knowledge for KBE systems, where attempts have been made to utilize this extension of UML in the knowledge engineering applications in Airbus design projects, for instance [Oldham et al. 1998]. In any case, in order to determine how much of an effect KBE has on the design process of optical backplane engineering, ratios comparing KBE vs. non-KBE techniques can be developed with respect to time, cost, and quality metrics.

Benefits of KBE

There are many benefits of using KBE over traditional methods. The time to produce an engineering artifact can be substantially reduced by eliminating tedious engineering duties through automated reasoning. The wisdom of the organization and key people can be retained and accessed by future generations of workers. Concurrent engineering can be accomplished when disparate departments can meld their work together by working in parallel, as opposed to sequentially, enabled by a KBE system intelligently updating files in the background according to its programmed logic. The entire design process will become fully documented, pulling together scattered annotations, designs, suggestions, and established heuristics.

Using new technologies, such as the Semantic Web, the KBE engineering system would enable designers to search repositories of parts globally, based on a sophisticated pre-recorded rule-base [Kopena, 2003]. CAD drawings can be automatically annotated, by intelligently capturing drawing intent and specifications – possibly even recommending new changes by taking into account overnight global technology shifts by a programmable logic advisor that sends intelligent agents into the Internet. The manufacturing and design process can be closely intertwined, avoiding costly redesign improvements. Consistency, and, thereby, quality control can be enhanced, throughout the engineering firm, when the designs are produced from the same set of rules from the knowledge base, avoiding source conflict [ICAD Release 7, 2004].

Although the initial KBE system may take longer to develop for an organization, it can evolve and adapt faster to new technology, demand, knowledge, and other market forces by maintaining its edge by capturing best practices and embedding them into a set of rules and ontologies usable by machine intelligence, systematically over time (Fig. 2.14). Thus, the traditional design process can go through many more iterations during the product development life-cycle using KBE techniques to refine the engineering process.

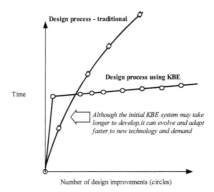

Fig. 2.14. KBE provides more design improvements over time.

KBE Product Model Captures Design Strategy

A product model can capture the design strategy to produce a particular product from a specification, based on rules or an ontology - which is a specification of a conceptualization [Miller, 2004]. Conceptualizations are generally provided by the domain expert and captured by the KBE system in the form of rules to form a knowledge base that can be built using ontologies. The ontology can describe the functional specifications of a part and its inter-relationships with other parts. As discussed later, frontier Semantic Web technology is expected to revolutionize this process by automating many engineering tasks by invoking rules applied to the Internet using intelligent agents. These rules can take many forms and be divided into multiple categories, which can be routinely updated by a domain expert through query sessions with the knowledge base.

KBE Rule Categories

Rule categories can include (1) rules that automatically formulate a part's specifications (2) rules that calculate engineering properties of a structure (3) rules that enable configuration selection according to limiting conditions (4) rules that optimize to improve on parameters such as cost, speed, and quality (5) rules that provide guidance on where and how to procure key information from the Internet and distributed databases (6) rules that provide analysis and second opinions (7) rules that impose the latest formats and standards to a design (8) rules that reveal design intent providing justification for an engineering decision and enabling design flexibility (9) rules that function as heuristics, providing fuzzy guidance on undefined problems. These rule categories can be expanded horizontally, as a domain expert fills in the details vertically per category [ICAD Release 7.0, 2004].

Knowledge Engineering Requires Meta-Modeling

KBE system is developed using Knowledge Engineering (KE) techniques that meta-model the knowledge into re-usable types that have lowered cost of development [Struder, 1998], usually at a higher abstraction level than software engineering constructs. Knowledge engineering does not mean mining the knowledge from a domain expert, but encompasses "methods and techniques for knowledge acquisition, modelling, representation and use of knowledge" [Schreiber et al., 1999]. Many knowledge engineering (KE) methodologies have been developed, such as CommonKADS [Schreiber et al., 1999], Protégé [Grosso, et al., 2000], and MIKE [Angele et al., 1998]. CommonKADS has been considered a mainstream language for knowledge engineering [**Abdullah et al.**, 2001]. Later, we will show how this knowledge modeling language forms the basis for developing a comprehensive task manager for the framework developed in this thesis. In fact, the CommonKADS approach plays a central role in the development strategy of the AIDF architecture, which we will describe in more detail.

Motivation for Knowledge Storage Method other than Rules

By a process of knowledge transfer, usually in the form of rules directly into the knowledge base, intelligent systems have been developed using knowledge engineering [Struder, 1998]. Despite the advantages of rapid knowledge elicitation and capture in the form of heuristics, formal and informal, disadvantage of this approach became apparent. As the size, scope, and type of knowledge grew, a meta-level method for capturing how the knowledge is related, linked, or connected together was needed [Schreiber et al., 1999]. Thus, an evolving knowledge base capable of being continuously updated and authenticated by multiple experts concurrently became a problem. As a result, a new modeling strategy promoting efficiency that reduced development costs based on ontologies emerged, which enabled development based on re-usable knowledge components that could be intimately woven together through semantic networks.

Knowledge stored as Ontologies exploitable by Semantic Web

Ontologies are simply specifications of a concept that have can be used to represent knowledge and their relationships [Gruber, 1993]. In knowledge modeling, ontologies define the domain-specific knowledge representation elements, including domain-dependent classes, relations, functions and objects [Gomez-Perez et al., 1999; Gomez-Perez et al., 2000]. Assisting in the design process, new Semantic Web languages, such as OWL, are enabling engineering components, for instance, to be "marked up" in a form that a computer can understand and intelligent agents can search and retrieve to [Kopena, 2003]. These methods can significantly improve the intelligent automation of reliability engineering by bringing to bear the fruit of engineering activity throughout the world for product development – virtually on demand – throughout the design process. In this way, domain knowledge can be stored, represented, and processed in knowledge-based systems connected to Web Services, allowing thousands of domain experts in the field to take part in the knowledge engineering process by updating their knowledge by virtual templates.

Knowledge stored as Problem Solving Method (PSM)

A Problem Solving Method (PSM) describes the reasoning-process in terms of a generic inference patterns at an abstract level [Grosso, et al., 2000]. PSMs have influenced the knowledge-engineering frameworks, including CommonKADS, Protégé, MIKE, Components of Expertise, EXCEPT, GDM and VITAL [Gomez-Perez et al., 2000]. It appears that a PSM decomposes the higher-level reasoning task into its constituent inferences. Then a PSM defines the types of knowledge that will be utilized by the inference steps outlined, as well as the control mechanisms and flow of knowledge among the inferences.

COMMONKADS APPROACH TO KNOWLEDGE ENGINEERING

CommonKADS: De Facto Standard for Knowledge Modeling

Ideal for developing a KBE system, the Common Knowledge Analysis and Design Support (KADS) approach to knowledge engineering has become the de facto standard for knowledge modeling [Abdullah et al., 2001]. A trend has been observed in the use of OO modeling by both developers and researcher to develop conceptual models for

KB systems influence by CommonKADS [Felfernig et al., 2000; Manjerres, 2002]. Prac-

titioners of MOKA, a methodology for knowledge based engineering, have based their

knowledge modeling on CommonKADS [Stokes, 2001]. In addition to providing stan-

dard methods for performing intricate analysis of knowledge intensive tasks, the knowl-

edge modeling structures knowledge engineering techniques and provides tools for cor-

porate knowledge management. CommonKADS introduces a full suite that supports

modeling of the knowledge and tasks, the intelligent agents that implement the tasks, and

the comprehensive design of the knowledge management system. CommonKADS has an

object-oriented development process using UML notations, making it compatible with

other software engineering projects. CommonKADS also includes graphical notations

for task decomposition, inference structures and domain schema generation [Schreiber et

al., 1999].

CommonKADS Knowledge Category Definitions

A knowledge model has three parts (Table 2.29), each capturing a related group

of knowledge structures, called a *knowledge category*, namely *Domain Knowledge, In-

ference Knowledge*, and *Task Knowledge* [Schreiber et al. , 1999].

Table 2.29. CommonKADS knowledge category definitions.

#	Knowledge Categories	Definition
1	Domain Knowledge	Composed of domain types, rules, and fact that specify the domain-specific knowledge and information that we want to talk define in an application, including relationships between knowledge types (i.e. ontologies); sometimes called a data model or object model
2	Inference Knowledge	Composed of basic inference mechanisms and roles that describe the basic inference steps using the domain knowledge. Inferences are best seen as the building blocks of the reasoning machine. In software engineering terms, the inferences represent the lowest level of functional decomposition.
3	Task Knowledge	Composed of task goals, decomposition, and control mechanisms that describe what goal(s) an application pursues, and how these goals can be realized through a decomposition into subtasks and (ultimately) inferences. Task knowledge is similar to the higher levels of functional decomposition in software engineering.

CommonKADS Analysis and Synthesis Task Decomposition

The knowledge intensive tasks is comprised of *analytic* and *synthetic* tasks (Table 2.30), which have appropriate sub-classifications, based on type of reasoning employed using the CommonKADS approach to KBE (Fig. 2.15). Analysis is defined as a type of super task requiring breakdown of information into constituent parts, enabling a focused investigation. Synthesis is defined as a type of super task requiring combination of parts or elements so as to form a whole, enabling a general symbiosis of parts.

Table 2.30. CommonKADS analysis and synthesis definitions.

Type of Knowledge Intensive Task	Description
Analysis	A type of super task requiring breakdown of information into constituent parts, enabling a focused investigation
Synthesis	A type of super task requiring combination of parts or elements so as to form a whole, enabling a general symbiosis of parts

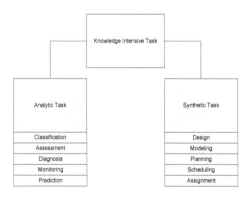

Fig. 2.15. CommonKADS hierarchy of tasks.

CommonKADS Analytic Tasks

There are five types of *analytic* tasks, namely classification, diagnosis, assessment, monitoring, and prediction assignment (Table 2.31). *Classification* is defined as an analysis task involving characterization of an object in terms of class. *Diagnosis* is defined as an analysis task involving determination of faults in the system. *Assessment* is defined as an analysis task involving characterization of a case in terms of a decision class. *Monitoring* is defined as an analysis task involving analysis of a system of a dynamic nature. *Prediction* is defined as an analysis task involving analysis of current system behavior to construct a description of the system state at some future point in time.

Table 2.31. CommonKADS analysis task types.

Analysis	Description
Classification	An analysis task involving characterization of an object in terms of class
Diagnosis	An analysis task involving determination of faults in the system
Assessment	An analysis task involving characterization of a case in terms of a decision class
Monitoring	An analysis task involving analysis of a system of a dynamic nature
Prediction	An analysis task involving analysis of current system behavior to construct a description of the system state at some future point in time.

CommonKADS Synthesis Tasks

There are five types of *synthetic* tasks, namely design, modeling, planning, scheduling, and assignment (Table 2.32). An example of a subtask of *design* is *configuration*. *Design* is defined as a synthetic task such that the system to be constructed is a physical artifact. *Assignment* is defined as a synthetic task such that a mapping between a set of objects is made. *Planning* is defined as a synthetic task such that physical object construction is done having time dependencies, giving a sequence of activities. *Scheduling* is defined as a synthetic task such that planned activities are allocated resources. *Modeling* is defined as a synthetic task such that an abstract description of a system is constructed.

Table 2.32. CommonKADS synthesis task types.

Synthesis	Description
Design	A synthetic task such that the system to be constructed is a physical artifact
Assignment	A synthetic task such that a mapping between a set of objects is made
Planning	A synthetic task such that physical object construction is done having time dependencies, giving a sequence of activities
Scheduling	A synthetic task such that planned activities are allocated resources
Modeling	A synthetic task such that an abstract description of a system is constructed

DESIGN PROCESS SUPPORT WITH INTELLIGENT AGENTS ON SEMANTIC WEB

Semantic Web as part of Web Services

The Semantic Web is an extension of the World Wide Web in which information is given well-defined meaning, better enabling computers and people to work together [Daconta et al., 2003]. Ontologies, being formal specifications of concepts, play an important role in the Semantic Web as a method for semantic representation of documents, so that the semantics can be used by Web applications and intelligent agents such as Web-searching software. The growth of the Internet has led to ontology language development exploiting the advantages of global knowledge acquisition using Web Services.

Web Ontology Language for Global Knowledge Acquisition

In 2001, the W3C created the Web-Ontology Working Group in order to develop a new ontology markup language for the Semantic Web, called the Web Ontology Language (OWL) [Miller, 2004]. Ontology mark-up languages such as DAML, developed in the U.S, and OIL, developed in Europe, are converging to a common standard OWL that can handle various knowledge representation tasks and reasoning tasks. Intelligent

agents could be programmed with both reusable and machine processible domain expertise enabling automatic assistance in the design process of optical backplane engineering,
while the governing logic and rules in the KB could be continuously updated and expanded.

Semantic Web Languages, Support, and Tools

Already some applications have been developed using a new language, such as
DamlJESSKB, that is able to search and retrieve hardware components already marked-
up in OWL using the Semantic Web [Kopena, 2003]. In addition to being able to export
ontologies developed in CmapTools to OWL, other tools have been developed for integration of a KB with the Semantic Web [Concept Map, 2006]. Developed at Stanford,
Protégé is a free, Java-based, open-source ontology editor and knowledge-based framework, closely tied to Jena, the language used for inference development by the Semantic
Web. Protégé-Frames implements a knowledge model compatible with Open Knowledge
Base Connectivity (OKBC), which is a DARPA sponsored application programming interface for accessing knowledge bases stored in knowledge representation systems, developed under the sponsorship of the High Performance Knowledge Base (HPKB) program. Knowledge mining for building distributed applications on the Internet can be
done using Protégé [Houlding, 2006]. We have considered these types of new developments in our architectural framework, so we can exploit authenticated knowledge repositories developed by domain experts on the Web. Authentication of remote knowledge
repositories will be discussed in validation. Four major stages in the development of the
web include the Web as a collection of accessible information, the Semantic Web making

possible mechanized semantic information processing, Web Services for distributed computing, and Semantic Web Services combining both for mechanized service discovery and execution [Fensel, 2004] (Fig. 2.16).

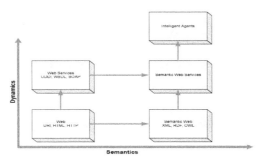

Fig. 2.16. Semantic Web, Intelligent Agents, and Web Services.

VALIDATION APPROACH

This section is only an overview of validation taken in this thesis, which is an important subject, which is handled in more detail with a dedicated chapter on validation with a corresponding expansion provided in Appendix A.

VALIDATION WITH RESPECT TO VERIFICATION

Defining Validation, Verification, and Requirement Specifications

The ultimate goal of *validation* is making sure that the "right" system is developed for the end-user, usually the customer, with respect to meeting *real-world needs*, whereas the goal of *verification* is making sure that the system was developed "right", with respect to meeting the engineering *requirement specifications* [Hoppe, 1993]. By assuring end-user formal and informal needs are clearly identified, hierarchical decomposition of system requirements significantly increase the likelihood of achieving design

75

success during validation and verification (Table 2.33). When the engineering requirement specifications actually meet the end-user real-world needs, a *validated and verified*
system can be deployed. In the case of software development, design and functional
specifications must be met for verification [Jayaswal and Patton, 2006].

Table 2.33. Validation and verification of requirement specifications.

Validation Terminology	Description
Requirement Specifications	Assures end-user formal and informal needs are clearly identified
Validation	Assures the "right" system is built to meet real-world end-user needs
Verification	Assures the system is built "right" to meet engineering specifications

Validation Subsumes Verification of Requirement Specifications

Validation usually subsumes verification, so that system requirements are checked
to meet engineering specifications *before* checking to make sure the system meets end-
user needs [Guedez, et al.2001]. Many subsystems can also be independently verified
and validated before subsequent system integration and final validation (Fig. 2.17).

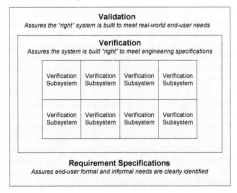

Fig. 2.17. Validation and verification of system requirements.

Improving On Time, Quality, Cost For KBE Development

In engineering, *timeliness*, *quality*, and *cost*, which have been widely accepted by engineers to be central performance metrics [Muirhead and Simon, 1999; Muirhead, 2004]. An ideal architectural framework should be able to improve on speed, cost, and time for KBE development which allows for faster development time, higher quality with pre-deployment and post-deployment validation strategy, and less expensive production for all KBE systems configured down the line. A validated, reconfigurable, and scalable architectural framework, having a high Return On Investment (ROI) for a KBE implementation, is able to deliver with respect to these three attributes.

THESIS APPROACH TO VALIDATION

In this thesis, we present a validated architectural framework that functions as a reconfigurable platform for design process automation with artificial intelligence leveraging intelligent agents on the Semantic Web. In order to achieve comprehensive validation, we researched the state-of-the-art standards, techniques, and methods employable by KBE SoS configured by the AIDF and developed a three-layer stratification.

Comprehensive Validation by Three-Layer Stratification

For comprehensive validation (Fig. 2.18), we *stratified* our approach into three layers: divisions, areas, and methodologies. First, we scoped four divisional targets. Then, we decomposed the divisions into their corresponding validation areas, each having respective methodologies comprised of employable standards, techniques, and meth-

ods. The *four divisional* (Table 2.34) targets comprised of multiple application areas cor-

responding to *software architecture* (Table 2.35), *design process* (Table 2.36), *artificial*

intelligence (Table 2.37), *and knowledge acquisition* (Table 2.38). The associated meth-

odologies (Table 2.39) for each area were explored and distilled, with the resulting state-

of-the-art standards, techniques, and methods identified and described in this chapter, re-

serving the detailed explanation in the validation chapter. The references for all the divi-

sions, areas, and methodologies are provided in the validation chapter.

Validation of Identified Target Areas for Automated Product Design Support based on Artificial Intelligence and Web Services			
AIDF Validation *Target I* **Software** **Architecture**	*AIDF Validation* *Target II* **Engineering Product** **Design Process**	*AIDF Validation* *Target III* **Artificial Intelligence** **Inference Mechanisms**	*AIDF Validation* *Target IV* **Global Knowledge** **Acquisition Process**
Areas addressed	*Areas addressed*	*Areas addressed*	*Areas addressed*
(A) Architecture development process (B) Architecture framework model (C) Architectural structural/dynamic model (D) Architectural process model	(A) Engineering product design phases (B) Prevailing engineering design methods currently available	(A) Automated Knowledge Analysis & Design Synthesis Approach (B) Knowledge-based Ontology Implementation (C) Knowledge-based Ontology Evaluation (D) Knowledge-based Weighted Rules Verification (E) Knowledge-based Algorithmic Methods	(A) Authentication of Remote Knowledge Repositories (B) Intelligent Agents on Semantic Web
Set of Global Standards and Techniques Identified			

Fig. 2.18. Validation of identified division target areas.

Table 2.34. High-level validation target divisions.

Validation Target Division
(I) SOFTWARE ARCHITECTURE
(II) PRODUCT DESIGN PROCESS
(III) ARTIFICIAL INTELLIGENCE
(IV) KNOWLEDGE ACQUISITION

Table 2.35. Validation areas for target division I: Software architecture.

Validation Target Area	Appendix
(A) Architecture development process	I-A
(B) Architecture framework model	I-B
(C) Architectural structural/dynamic model	I-C
(D) Architectural process model	I-D

Table 2.36. Validation areas for target division II: Design process.

Validation Target Area
(A) Systems engineering *product design phases*
(B) Prevailing *engineering design methods*

Table 2.37. Validation areas for target division III: Artificial intelligence.

Validation Target Area
(A) Automated Knowledge Analysis and Design Synthesis Approach
(B) Knowledge-based Ontology Implementation
(C) Knowledge-based Ontology Evaluation
(D) Knowledge-based Weighted Rules Verification
(E) Knowledge-based Algorithmic Methods

Table 2.38. Validation areas for target division IV: Knowledge acquisition.

Validation Target Area
(A) Authentication of Remote Knowledge Repositories
(B) Intelligent Agents on Semantic Web

Table 2.39. Validation targets, application areas, standards, techniques, methods.

Target	Application Areas	Validation implementation standards, techniques, methods
(I) Software Architecture	(A) Architecture development process	*(1) ANSI/IEEE Std 1471 (2) Clinger-Cohen Act (3) OMG-MDA (4) First International Workshop on IT Architectures for Software Systems in 1995, (5) Shaw-Carnegie Mellon architectural categorizations (6) ATAM for architectural trade-off analysis, (7) CBAM for architectural cost assessment, (8) Acclaro Design for Six Sigma (DFSS) implementation, (9) Telelogic TAU implementation of UML and SysML, (10) Rational Unified Process (11) Architecture-Driven Software Construction*
	(B) Architecture framework model	*(1) AIDF implementation/Acclaro Design for Six Sigma, (2) Failure Mode and Effects Analysis (FMEA), (3)Design Structure Matrix (DSM) dependency resolution, (4) Axiomatic Design Theory (ADT) risk mitigation, (5) Quality Function Deployment (QFD) (6) Generic Architecture for Upgradeable Real-Time Dependable Systems (GAURDS) validation framework (7) Survey of 25 decision support tools and 10 frameworks*
	(C) Architectural structural/dynamic model	*(1) High Level Integrated Design Environment (HIDE) for dependability (2) Telelogic TAU SysML implementation/code validation/verification*
	(D) Architectural process model	*(1) Axiomatic V-Model mapping process (2) National Academy of Engineering 2001 Report on Approaches to Engineering Design*
(II) Design Process	(A) Engineering product design phase	*Methodology by Pahl and Beitz*
	(B) Prevailing engineering design methods currently available	*(1) Axiomatic Design Theory (ADT), (2) Theory of Inventive Problem Solving (TRIZ), (3) Hierarchical Multi-layer Design (MLH), (4) Quality Function Deployment (5) Design Structure Matrix (DSM) (6) Fault-Tree-Analysis (FTA) (7) Failure-Mode and Effects Analysis (8) Reliability Block Diagram (RBD) Analysis (9)Technology Risk Factor (TRF) Assessment (10) Entropy (ETP) Analysis, and (11) Case Study: Optical Backplane Engineering (OPT) Domain*
(III) Artificial Intelligence inference Mechanisms	(A) Automated Knowledge Analysis and Design Synthesis Approach	*(1) the de-facto standard for knowledge modeling as defined by Common Knowledge Analysis and Design Synthesis (CommonKADS) Approach, and the (2) the European Union Esprit II VALID project exploring the state-of-the-art for validation of KB systems*
	(B) Knowledge-based Ontology Implementation	*(1) Concept Map tools (CmapTools) for domain expert knowledge capture with ontologies, (2) Concept Map tools (CmapTools) Methodology to connect to global knowledge repositories on Semantic Web by demonstrating the exportability of ontology constructs into XML and Web Ontology Language (OWL) for use by intelligent agents*
	(C) Knowledge-based Ontology Evaluation	Mathematical assessment of ontology quality and quantity in terms of *breadth, fan-out,* and *tangledness*
	(D) Knowledge-based Weighted Rules Verification	*(1) verification with HP Labs Jena coding for inference engine, (2) code evaluation of Java Expert System Shell (JESS), (3) weighted rules verification with Matlab Fuzzy Logic Toolbox*
	(E) Knowledge-based Algorithmic Methods	*(1) domain rule support (2) predicate logic support (3) algorithmic reasoning support (4) fuzzy logic support (5) neural network support (6) genetic algorithm support (7) Conant transmission support (8) Calibrated Bayesian Support (9) Data Mining Support*
(IV) Global Knowledge Acquisition Process	(A) Authentication of Remote Knowledge Repositories	*Cluster-on-demand architecture (Coda)*
	(B) Intelligent Agents on Semantic Web	*DAMLJESSKB* and its evaluation and verification methods

SUMMARY

In this chapter, we have provided a literature review of research for a reconfigurable architectural framework application developed for reliability engineering automation leveraging Knowledge-Based-Engineering (KBE) techniques networked to domain experts using intelligent agents. We have show that broad topics integrated by such a SoS application include many interrelated fields, such as systems engineering, software and product engineering design, software and system architecture development, risk mitigation, reliability engineering, intelligent agents, human factors engineering, and knowledge-based engineering utilizing quality design and artificial intelligence algorithmic techniques. An in-depth review of these topics has provided a foundation for intelligently automating the engineering design process for reliable product design impacting the fields of computer engineering, system-of-systems, and architecture development. We have shown that intellectual control of a KBE system-of-systems, characterized by a large-scale system capable of managing a set of complex, distributed systems, can be achieved by adapting a reconfigurable and scalable architectural framework for reliability engineering automation.

CHAPTER III: ARTIFICIAL INTELLIGENCE DESIGN FRAMEWORK

OVERVIEW

In this chapter, we introduce and define the AIDF, providing its scope and impact. We provide a validation overview for the architecture framework. We introduce the primary views of the framework in terms of architecture-driven software engineering, the AIDF itself, AIDF internal and external dataflow, AIDF design process with management gates, AIDF interlacing dual engine block configuration supporting reliability, and AIDF reliability engine dynamics with Web Services We provide a discussion of the chapter contents, before describing the AIDF input, processing, and output block that operate as the Knowledge Assimilation Engine (KAE), Knowledge Correlation Engine (KCE), and the Knowledge Justification Engine (KJE), respectively. We describe each engine block in detail, providing snapshots of the implementation with Acclaro Design for Six Sigma. We introduce the CommonKADS approach as a basis for Knowledge Modeling in order to develop the framework's artificial intelligence task manager. We describe the operations in each block, followed by an introduction of the three stages of the AIDF corresponding to the three design phases. The modular engine operations in each of the three AIDF design stages are examined with a design matrix. A discussion of the strategy used for development of the architecture is provided. The details for each of the twenty modules in operation during reliability engineering automation are described in Appendix B. We introduce a standard that assists with modularity with a final summary.

SOFTWARE ARCHITECTURE FRAMEWORK SCOPE AND IMPACT

Architecture Framework Developed to Configure KBE Systems for Reliability

Our goal is to provide an architectural framework that can configure KBE systems managing complex systems that support design risk mitigation during reliability engineering. We use the architecture-driven software engineering process based on the axiomatic V-model to produce a KBE SoS application for optical backplane engineering. The *Artificial Intelligence Design Framework (AIDF)* (Fig. 3.2) is divided into three blocks: (1) Knowledge Assimilation Engine (KAE) that gathers project data and networked knowledge using intelligent agents (2) Knowledge Correlation Engine (KCE) that processes this knowledge for design risk mitigation for reliability engineering (3) and a Knowledge Justification Engine (KJE) that outputs justifiable design recommendations based on the processing using a model-view-controller approach.

The framework is reconfigurable and scalable, as well as extensible, to include networked knowledge repositories developed by remote domain experts concurrently. Additionally, it has the capability of adapting a set of KBE systems for future software development and deployment on demand using model-driven architecture approach, based on verified requirement specifications meeting real-world needs as defined by validation. Furthermore, the framework has a methodology for validation before and after deployment.

In advance, we introduce the following diagrams that we will refer to throughout the thesis: the *Architecture-driven software engineering process* (Fig. 3.1), the *Artificial Intelligence Design Framework* (AIDF) (Fig. 3.2), the *AIDF architecture framework op-*

83

eration (Fig 3.3) managing the flow, and the AIDF three-stage *design process model* (Fig. 3.4) having a corresponding stage for each design phase.

The scope of the thesis is established (Fig. 3.1) according to the architecture-driven software engineering process mapping the functional requirements to object-oriented code, which ultimately forms a KBE SoS application configured to a domain, when seamlessly integrated. Since our focus in the case study is on design risk mitigation for reliability engineering, we will show the operations of the Reliability (RBD) module and its interlacing connections in the *AIDF dual engine block* (Fig. 3.5), whose *reliability engineering automation during embodiment design* is detailed in the case study chapter (Fig. 3.6). All of these diagrams will be elaborated on in detail in the proceeding chapters.

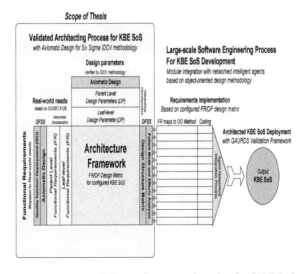

Fig. 3.1. Architecture-driven software engineering for KBE SoS.

Artificial Intelligence Design Framework (AIDF) Architecture Model

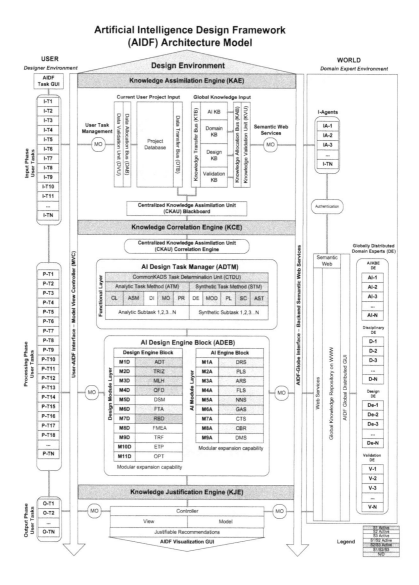

Fig. 3.2. AIDF architectural framework model detail.

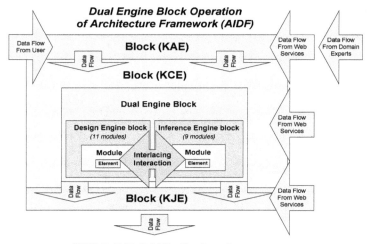

**KBE SoS Reliability Engineering
Recommendations Output**
*Based on AIDF design matrix requirements
configured for case study (per stage)*

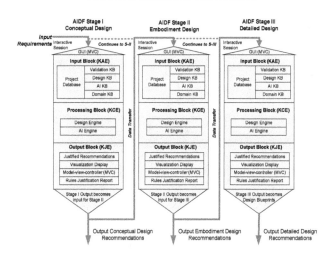

Fig. 3.3. Internal and external data flow of AIDF.

86

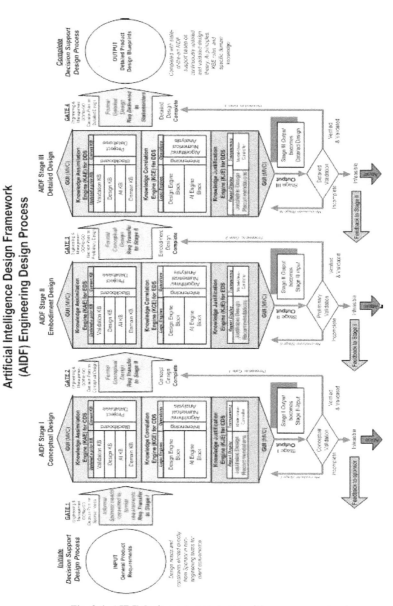

Fig. 3.4. AIDF design process stages with management gates.

Interlacing Connections between Dual Engine Block
for Reliability Automation of Configured Case Study

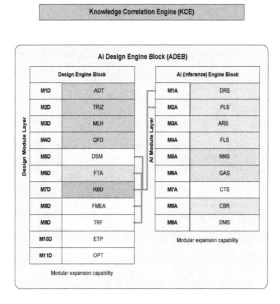

Dual Engine Block Operation
of Architecture Framework (AIDF)

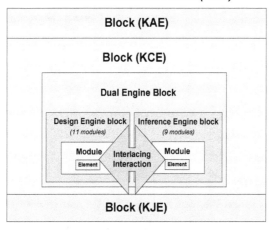

Fig. 3.5. Dual engine block interlacing connections for configured case study.

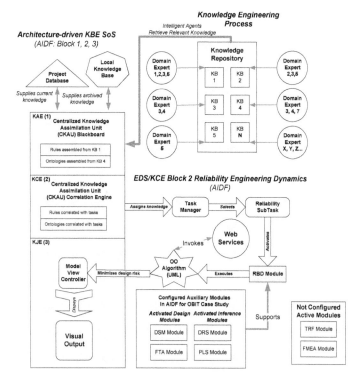

Fig. 3.6. Reliability engineering dynamics defined by AIDF EDS/KCE.

Architecture Framework Impacting Three Fields

Managing complex systems has become a major challenge in achieving quality in large-scale, dynamic systems having millions of interacting components. As the number, components, and relations between components increase, the design complexity of a system may exceed intellectual control and become an entangling problem. The architectural framework provides a structure for automatically managing domain-specific product engineering knowledge for design support connected to networked knowledge bases is needed. Furthermore, an architectural framework that fulfills this need should also be able to provide an overarching structure for orderly development and integration of complex systems interacting with networked knowledge repositories available on the Web, as well as local knowledge bases. Such an architectural framework that provides such intellectual control over complex interacting systems draws from and impacts three broad fields (Fig. 3.7): computer engineering, system-of-systems and architecture development.

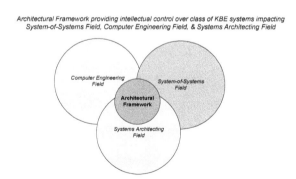

Fig. 3.7. AIDF provides intellectual control and impacts three fields.

Rationale for Architectural Framework Development

The *rationale* for developing an architectural framework is to ensure that intellectual control of a *System-of-Systems* (SoS) engineering problem spanning many fields is
provided. Thus, the *framework* should function as a reconfigurable and scalable platform
to manage large-scale, distributed, agent-based complex systems automatically interacting with a KBE system. An appropriate architectural framework must be developed that
provides a structure for an automated design environment KBE system comprised of a set
of *complex systems* capable of brute-force *design* calculations, as well as algorithmic *artificial intelligence* processes, by accessing local and remote knowledge repositories with
intelligent agents based on the most current technology. These complex systems are *dy-
namically changing* daily, due to the very nature of continuous knowledge updating and
retrieval using Web Services.

Purpose of Configured Architecture Framework for Case Study Instantiation

Hence, the *overarching purpose* of the *configured* architecture for a particular instantiation of a KBE system, e.g. optical backplane engineering case study, is to provide a
structure for managing such complex dynamic systems. In this way, the architecture es-
tablishes a structure to guide software and systems engineering development for large-
scale, distributed systems by melding together an unwieldy set of distinct fields into a
unified whole. Furthermore, the high-level, interdisciplinary nature of systems-of systems
paradigm is critical for understanding how an architectural framework fits into the general architectural description for a system-of-systems paradigm.

SOFTWARE ARCHITECTURE DEVELOPMENT FOR SYSTEM-OF-SYSTEMS

Need for Intellectual Control over Development of Complex Systems

Managing complex systems has become a major challenge in achieving quality in large-scale, dynamic systems having millions of interacting components. As the number, components, and relations between components increase, the design complexity of a system may exceed intellectual control and become an entangling problem. A structure for automatically managing domain-specific product engineering knowledge for design support connected to networked knowledge bases is needed. Furthermore, an architectural framework that fulfills this need should also be able to provide an overarching structure for orderly development and integration of complex systems, including but not limited to architecture development, design process, artificial intelligence mechanisms, and agent-based knowledge acquisition techniques.

Defining Architecture Scope in Systems Engineering

The processes associated of the Tufts' Systems Engineering Process Model (Fig. 3.8) product lifecycle model are based on the Integrated Capability Maturity Model (CMMI) [INCOSE, 2006a]. It should be noted that, with respect to the scope of the thesis, we are focusing on architecture development and validation for a KBE system, which actually comprises only one of eight distinct aspects of the systems engineering product life cycle. In this case, we are showing that the high-level architecture developed in this thesis can develop configured KBE systems as a "product" line. The architecture in this diagram can also refer to that developed for the structure of components by engineers, a process itself that can be automated.

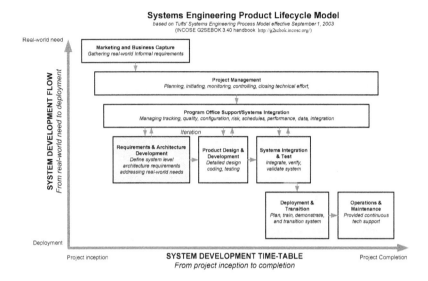

Fig. 3.8. Tufts' Systems Engineering Process Model.

Architectural Framework for Intellectual Control of KBE System-of-Systems

Development of a suitable architecture framework for a class of KBE system sp e-
cialized for design support based on automated artificial intelligence leveraging the S e-
mantic Web is a *system-of-systems level problem,* requiring a suitable paradigm (Fig.
3.9). The level of detail involved in gaining intellectual control over a set of complex
systems managed by the KBE system is great, which exceeds architecture for a well-
defined system, as in systems engineering. For instance, orchestration of thousands of
intelligent agents responsible for retrieving knowledge and hundreds of domain experts
updating that knowledge is a formidable task, especially when introducing authentication
and security issues at the common knowledge repository interfaces.

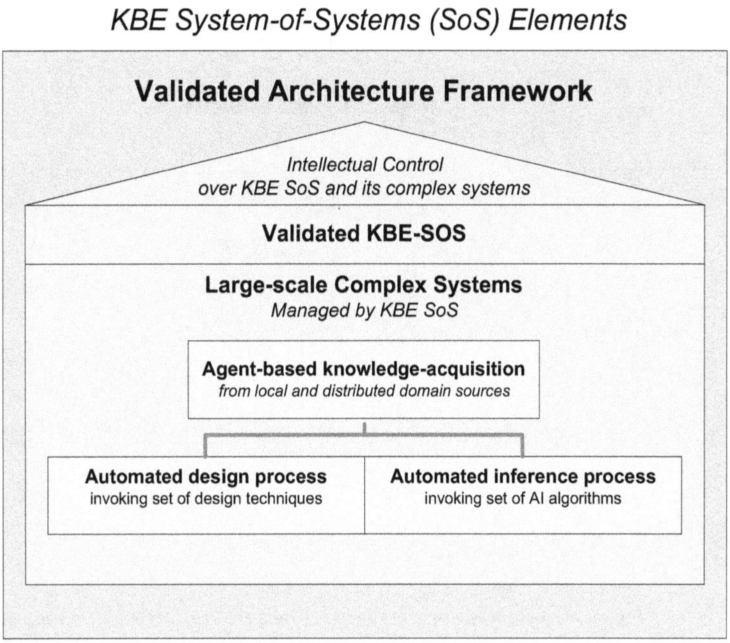

KBE System-of-Systems (SoS) Elements

Fig. 3.9. System-of-Systems Paradigm for Artificial Intelligence Design Support.

Axiomatic Functional Requirements Specification for Architectural Framework

In systems engineering, architecture development plays a pivotal role understanding the needs of the end-user, usually the customer, and translating it into a language engineers can follow, usually in the form of blueprints [Maier and Rechtin, 2000]. These blueprints are essentially a compendium of detailed requirements specifications. One leading method for design endorsed by the National Academy of Engineering [NRC, 2002] is axiomatic design analysis based on the Axiomatic Design Theory developed at Massachusetts Institute of Technology (MIT). The process model prescribed by axiomatic design is the V-model [Suh, 2001] for architectural framework construction (Fig.

3.10). This framework enables the architect to build and achieve intellectual control over large-scale architectures for complex, object-oriented systems.

Axiomatic Design of Architecture Framework

Fig. 3.10. Axiomatic process for architecture framework development.

Implementation of High-Level Architecture Framework

An architectural framework for systems engineering capturing end-user real-world needs can be implemented directly after requirements consultation, as a basis for rapid-prototyping using the architectural development tool Acclaro Design for Six Sigma (DFSS) [El Haik, 2004]. For large-scale applications, a reconfigurable and scalable plat-form provides features for front-end validation and verification using DFSS techniques before software engineering. The central product of Acclaro DFSS is a large-scale design matrix displaying full-scale architecture requirements in the form of a framework (Fig. 3.11), whose functional requirements can be mapped to object-oriented methods for further detailing.

Architecture-driven software engineering
for KBE System-of-Systems (SoS)

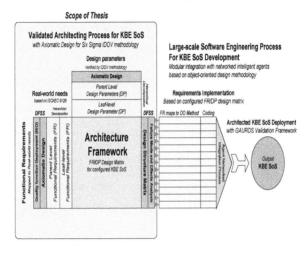

Fig. 3.11. Architecture-driven software engineering for KBE SoS.

VALIDATION STRATEGY OF AIDF ARCHITECTURE FRAMEWORK

Architecture Validation Overview

We developed the AIDF architectural framework as a platform that takes into account many concerns, including architecture, systems engineering, design, as well as validation. Specifically for validation, a Synergistic Validation Methodology (SVM) was developed to address four aspects, or divisions, of the AIDF: (I) Architecture Development, (II) Design Process, (III) Artificial Intelligence (IV) Knowledge Acquisition. Each division is comprised of multiple areas of validation. Each area is further broken down into methodologies in terms of standards, techniques, and methods applied. We recommend a process for software engineering and systems engineering teams to apply these techniques at various stages of the architecture development and KBE system configuration stages in the *ten-step, five-tier* cascading validation process we developed. This *validation process* provides guidance to software and systems engineers on when and how to apply the various global standards and methodologies we identified and bundled together into areas (See Validation Chapter).

Features Matching Benefits

When a verified system's *features* correspond to its validated *benefits*, a real-world system can be developed that addresses not only rigorous functional requirements defining a system, but also stipulated application needs stated by the project sponsor, who is usually the customer requesting a domain-specific KBE system. The processes of validation, verification, evaluation, and testing form an integral part of this process, in order to ensure that the final system actually meets the end-user requirements.

Avoiding Requirements Trap and Requirements Creep

Various front-end validation techniques are applied to the AIDF to prevent *implementation trap*, a condition where original functional requirements are not addressed or lost during downstream design decisions made during software engineering implementation [Albin, 2003]. Furthermore, *requirements creep,* a condition of late-stage functional add-ons defining KBE system needs, can not only be avoided, but flexibly addressed by the AIDF due to its modular configurability for each domain of application. This front-end approach is especially useful for validation to meet user expectations by addressing functional requirements that are traceable directly to top-level domain-specific customer needs to avoid implementation trap and requirements creep.

Systematic Validation of a Series of Configured KBE SoS Applications

For front-end validation, the Identify-Define-Optimize-Verify (IDOV) approach

using analysis techniques such as Axiomatic Design Theory (ADT), Design Structure

Matrix (DSM), Quality Function Deployment (QFD), Failure Mode and Effects Analysis

(FMEA), can be conducted using Acclaro DFSS, a tool also recently acquired by General

Dynamics architecture for development for large-scale engineering projects [Axiomatic

Design, 2005]. For post-deployment validation of a real-time system, the GAURDS

validation framework can be applied to the front-end validated architecture framework

already done by Acclaro DFSS. Combining the *software architecture* validation with the

other three divisions, i.e. *design process, artificial intelligence, and knowledge acquisi-*

tion, multiple KBE SoS applications can be produced as output of the architecture

framework, configured based on end-user real-world need in multiple domains (Fig.

3.12). Our case study represents one of the possible KBE SoS application configurations

possible based on the AIDF framework.

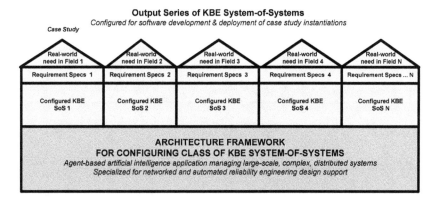

Fig. 3.12. Architectural framework capable of configuring class of KBE systems.

Primary Validation Target is the Software Architecture

Each validation division target can be decomposed into areas, each having respective methodologies comprised of employable standards, techniques, and methods. Hence, for the sake of comprehensive validation, we divided the comprehensive approach to validation by scoping four fundamental targets: *(I) Software Architecture, (II) Design Process, (III) Artificial Intelligence Inference Mechanisms,* and *(IV) Global Knowledge Acquisition Process.* Although much work on validation is done addressing as many areas as possible associated with the AIDF, the primary validation target is the software architecture (Fig. 3.13), since the AIDF itself is an architectural framework. Thus, validation of the implementation concentrates primarily on the framework. However, for comprehensiveness, the other three validation targets areas are also addressed in order to develop a synergistic methodology to configure a class of KBE systems having AI and Semantic Web elements based on the AIDF architectural framework platform.

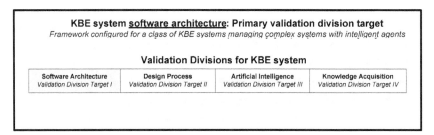

Fig. 3.13. Software architecture division is primary validation target.

Pre-deployment Validation Strengthened by Post Deployment Validation

We have separated our overall software architecture validation strategy into two stages (Fig. 3.14): (1) Pre-deployment and (2) Post-deployment of a KBE System-of-System (SoS). In pre-deployment we concentrate our efforts on front-end validation using Design for Six Sigma (DFSS) IDOV methodology [IDOV, 2006]. In post-deployment, we recommend the post-deployment application of a real-time architectural validation strategy for full-scale, continuous, and industry-grade validation of the front-end validated system using Generic Architecture for Real-time Upgradeable Dependable Systems (GAURDS) validation framework [Powell et al. 1999].

Software architecture validation strategy for KBE system	
KBE System Pre-deployment	Design for Six Sigma IDOV Methodology Front-end Architectural Validation Approach
KBE System Post-deployment	GAURDS Validation Framework Real-time Architectural Validation Approach

Fig. 3.14. Pre-deployment and post-deployment validation strategy.

Design for Six Sigma Contrasted with Traditional Six Sigma

When applying Design for Six Sigma (DFSS), a distinction should be made for its application to two different cases (1) already existing, or (2) to an unprecedented product or process. If it already exists but needs improvement, the DMAIC methodology, which stands for *Define, Measure, Analyze, Improve, control* is recommended in DFSS [Wortman, 2001]. DMAIC Six Sigma usually occurs *after* initial system or product design has been completed. However, if the product has never been developed before and needs quality control and front-end validation, the IDOV methodology is a recommended approach for reliability engineering, which stands for *Identify, Design, Optimize, and Verify*. Contrasted with this is the traditional DMAIC Six Sigma process, as it is usually practiced (Table 3.1).

Table 3.1. Design for Six Sigma IDOV methodology for front-end validation.

Step	Purpose	Method
Identify	Improve architecture by ensuring that functional requirements actually meet the original *end-user* needs	Identify real-world needs using method such as Quality Function Deployment (QFD) based on ISO/IEC 9126 defining standard features for end-user
Design	Improve architecture by ensuring that proper risk mitigation and reliability engineering methods have been applied in the modules	Design for systems level and component level risk mitigation by considering techniques like probability risk assessment (PRA) with Fault-Tree Analysis (FTA) and Reliability Block Diagram (RBD), reliability engineering using methods such as axiomatic design theory (ADT), Failure Mode and Effects Analysis (FMEA)
Optimize	Improve architecture by ensuring that efficiency is introduced into design	Optimize the configuration to ensure that dependencies are resolve using method such as Design Structure Matrix (DSM) and Theory of Inventive Problem Solving (TRIZ)
Verify	Improve architecture by ensuring that the architected solution actually meets the original *engineering* specifications	Verify that all the stated functional requirements are actually being met explicitly by the design parameters of the system by applying methods such as Axiomatic Design Theory (ADT)

GAURDS Validation Framework

The GUARDS validation strategy [GUARDS, 1997] considers both short-term and long-term objectives. This includes the validation of the design principles of the architecture and the validation of instances of the architecture for configured requirements, respectively. The validation environment that supports the strategy is depicted in the GAURDS diagram (Fig 3.15), which illustrates also the relationship between the components and their interactions with the architecture development environment. The main validation components in GAURDS depicted in the diagram are: (1) formal verification, (2) model-based evaluation, (3) fault injection, and (4) the methodology and the supporting toolset for schedulability analysis. In the case of fault injection, which is carried out on KBE prototypes, complements the other validation components by providing means for assessing the validity of the necessary assumptions made by the formal verification task. Also, it I useful in estimating the coverage parameters included in the analytical models for dependability evaluation.

AIDF Generic Architecture for Real-time Upgradeable Dependable Systems (GAURDS) Validation Framework

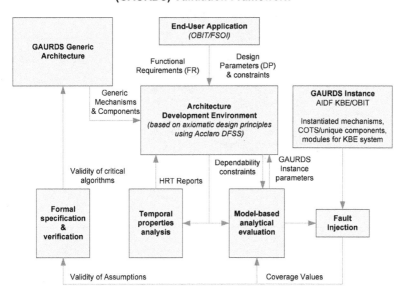

Fig. 3.15. GAURDS validation framework for post-deployment validation.

Summary for Architecture Validation

In order to validate the primary validation target of the AIDF involving *Software Architecture,* four areas were identified (Table 3.2) that addressed the (A) *architecture development process*, (B) *architecture framework model*, (C) *architectural structural/dynamic model*, (D) and *architectural process model.*

Table 3.2. Software architecture validation: Standards, techniques, methods.

Division	Application Areas	Validation Methodology standards, techniques, and methods
(I) Software Architecture	(A) Architecture development process	*(1) ANSI/IEEE Std 1471 (2) Clinger-Cohen Act (3) OMG-MDA (4) First International Workshop on IT Architectures for Software Systems in 1995, (5) Shaw-Carnegie Mellon architectural categorizations (6) ATAM for architectural trade-off analysis, (7) CBAM for architectural cost assessment, (8) Acclaro Design for Six Sigma (DFSS) implementation, (9) Telelogic TAU implementation of UML and SysML, (10) Rational Unified Process (11) Architecture-Driven Software Construction*
	(B) Architecture framework model	*(1) AIDF implementation/Acclaro Design for Six Sigma, (2) Failure Mode and Effects Analysis (FMEA), (3)Design Structure Matrix (DSM) dependency resolution, (4) Axiomatic Design Theory (ADT) risk mitigation, (5) Quality Function Deployment (QFD) (6) Generic Architecture for Upgradeable Real-Time Dependable Systems (GAURDS) validation framework (7) Survey of 25 decision support tools and 10 frameworks*
	(C) Architectural structural/dynamic model	*(1) High Level Integrated Design Environment (HIDE) for dependability* *(2) Telelogic TAU SysML implementation/code validation/verification*
	(D) Architectural process model	*(1) Axiomatic V-Model mapping process* *(2) National Academy of Engineering 2001 Report on Approaches to Engineering Design*

DISCUSSION

Acclaro DFSS used for Specification of Expandable Modules

We used *Acclaro Design for Six Sigma (DFSS)*, an architecture development tool used by General Dynamics for large-scale engineering projects. Design parameters fulfilling the functional requirements, constraints, and twenty possible modules supporting utilizable for automated product design support were identified and developed for suitability with the case study on optical backplane engineering. We used the architecture design tool to develop front-end validation of the AIDF by applying Axiomatic Design Theory for design risk mitigation by applying the Independence Axiom and Information Axiom for complexity reduction. Furthermore, Acclaro DFSS enabled advanced evaluation of potential problems encountered by a KBE system developed with the AIDF using

axiomatic design theory (ADT), Failure Mode and Effects Analysis (FMEA), Quality

Function Deployment (QFD), and the Design Structure Matrix (DSM).

Application of Framework to Optical Backplane Engineering

We were able to configure and validate a KBE system configured specifically for

optical backplane engineering by verifying high-performance architectural development

by decoupling techniques to eliminate design dependencies for the architectural compo-

nents and connectors. Many of the available technologies are shown in Appendix B,

which corresponds to the modules of the AIDF actually interacting for design automation

and inferencing. We were confident with the validation methodology of Acclaro DFSS,

considering that Axiomatic Design Theory is endorsed by the *National Academy for En-

gineering* in *Approaches to Engineering Design* annual report in 2001 on the state-of-the-

art for engineering design process, where Suh's theory enabling front-end validation re-

ceived the highest rating possible for the design category involving selection based on

alternative design [NRC, 2006].

Application of GAURDS to Validation

For further validation of a real-time KBE system launched by the AIDF, we have

adapted the safety-critical Generic Architecture for Upgradeable Real-time Dependable

Systems (GAURDS) validation framework used for nuclear submarines, space systems,

and railways for the AIDF itself. GAURDS allows functional architecture development

for real-time systems to be validated for integrity using commercial-off-the-shelf (COTS) components [Powell, 1999]. This feature of GAURDS is advantageous for the AIDF, considering that we have an architecture that is designed to leverage the most current technologies available for each of its twenty modules using COTS software. We will show in later chapters how the pre-validated AIDF itself fits into GAURDS validation framework used by safety-critical systems such as nuclear submarines, space systems, and railways. Furthermore, we will introduce in the SVM chapter how the AIDF leverages the High-level Integrated Design Environment (HIDE) methodology to validate the structural model using UML and SysML, a systems engineering language that has just emerged in OMG in May, 2006 [HIDE, 1999]. These areas of validation represent a few of the areas of validation identified and evaluated, where a more detailed treatment of this subject is done in the chapter on synergistic validation methodology.

Preliminary Research as a Foundation for Implementation

Before implementing and validating the AIDF, we conducted preliminary research which included a survey of over thirty decision support and artificial intelligence tools commercially available, in addition to dozens of frameworks for conceptual design [Wang et al., 2001]. The development of the AIDF is introduced in terms of the four phases of architecting, identified as *Predesign Phase, Domain analysis phase, Schematic Design Phase, Design Development Phase* [Albin, 2003]. Although outside the scope of this thesis, sometimes there is a *Build Phase* which involves project documentation, staffing and contracting, construction, and post-construction. Full-scale software development generally begins during or after the Build Phase. During the Predesign Phase, we were concerned mainly with understanding the context in which the AIDF application would exist; which was identified as a KBE system launching platform that provided automated decision support to design engineers. During the domain analysis phase, we scrutinized the AIDF application's domain-specific requirements for broad impact; which was identified as product design for engineering artifacts, in particular optical backplane design. During the schematic design phase, we began to develop the AIDF structure based on accepted standards and methodologies, defining the modules housed in the dual engine block; identified as 20 modules divided into the AI engine and design engine. During the design development phase, we implemented and validated the AIDF using Acclaro DFSS using front-end validation techniques; which was accomplished using the Synergistic Validation Methodology.

Development of the AIDF Architectural Framework Design Approach

For the development of the AIDF architectural framework, we have leveraged the widely accepted state-of-the-art methodology for KBE development using Common Knowledge Analysis and Design Support (KADS), the dominating industrial design process definitions by Pahl and Beitz [Pahl and Beitz, 1988], and the frontier of Web Services technology using W3C Semantic Web technology. For synthesis and analysis decision making methodology for the AIDF, we have selected the CommonKADS approach [Schreiber et al., 1999] to knowledge-based engineering. For the application of CommonKADS to the design process, we have based our definition of the three stages of design, namely conceptual, embodiment, and detailed design, on Pahl and Beitz. For the authenticated integration of the AIDF to remote knowledge depositories accessible by intelligent agents, we introduce the frontier of Semantic Web technology [W3C, 2006].

Modular Architecture Approach Having Three Distinct Stages and Blocks

Systems engineers, intelligent agents, domain experts, and knowledge engineers are continuously involved in eliciting, storing, validating, and updating *a prior* engineering theory and industrial best practices. Both deductive and inductive inference algorithms, as well as numerical analysis calculations, are processed by the design and AI (inference) engines. The architecture follows a modular architecture approach housing multiple modules capable of various lines of reasoning that provide analyses with justifiable recommendations. The AIDF is divided into three design stages, supporting conceptual, embodiment, and detailed [Pahl and Beitz, 1988]. Each stage is further divided into three main functional blocks, namely input, processing, and output block, referred to as the Knowledge Assimilation Engine (KAE), the Knowledge Correlation Engine (KCE), and the Knowledge Justification Engine (KJE), respectively.

Architectural Description based on Carnegie-Mellon School of Thought

Architectural description of the AIDF is categorized into four distinct models: *process, framework, structural,* and *dynamic* model [Shaw, 1996]. The AIDF framework model emphasized in this paper is implemented by Acclaro Design for Six Sigma (DFSS) software. The framework model is integrated with the structural and dynamic model using the Axiomatic Design process V-model [Suh, 2001]. The structural, dynamic, and functional models are developed using object-oriented modeling languages such as UML, SysML [INCOSE, 2006a] and MML [ESPRIT, 2000a]. The optical backplane engineering case study concentrates on Free-Space Optical Interconnect (FSOI) to demonstrate AIDF effectiveness with a prototype implementation for design risk mitigation [Kirk,

2003; Ayliffe, 1998]. The AIDF architecture utilizes architectural standards terminology of ANSI/IEEE Std. 1471-2000 [IEEE-1471, 2006] and model-driven architecture (MDA) [OMG, 2006] concepts to promote interoperability and stability at the highest levels of abstraction.

Models for Blocks and Stages in Chapter

In this chapter, we primarily focus on architectural models defining the AIDF, with some snapshots from Acclaro DFSS and other domain tools, such as CmapTools for ontology development. An overview of the AIDF model is introduced to show the conceptual approach to product design by focusing predominantly on the framework model, followed by the AIDF design process to show how the model supports product development, such as OBIT/FSOI design. The AIDF model is explained in terms of its three stages, where each stage is further subdivided into three blocks, having the functions of input, processing, and output (Fig. 3.16).

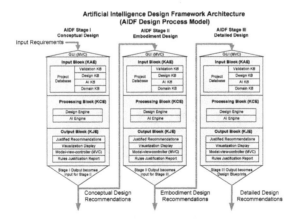

Fig. 3.16. Block subdivisions.

AIDF DESIGN STAGES SUBDIVIDED INTO FUNCTIONAL BLOCKS

The AIDF is divided into three design stages, following the standard phases of the design process. Following the model-driven architecture approach, each AIDF design stage is further subdivided into functional blocks (Fig. 3.17), detailed by the *Input Block (KAE), Processing Block (KCE),* and *Output Block (KJE)*, which are responsible for maintaining project and domain knowledge, processing data in the dual engine block, and displaying justifiable and traceable recommendations, respectively. The drill-down perspectives are (1) Framework-level, (2) Block level (3) Engine-level, (4) Module-level, and (5) Element-level (Fig. 3.18). Hence, using tools such as Acclaro Design for Six Sigma we are able to drill down all the way to elemental level from high-level framework level where all the functional requirements are updated, according to the end-user needs. The benefit of this process is traceability of the actual coding to the original requirements. We will detail element level interaction in the case study chapter.

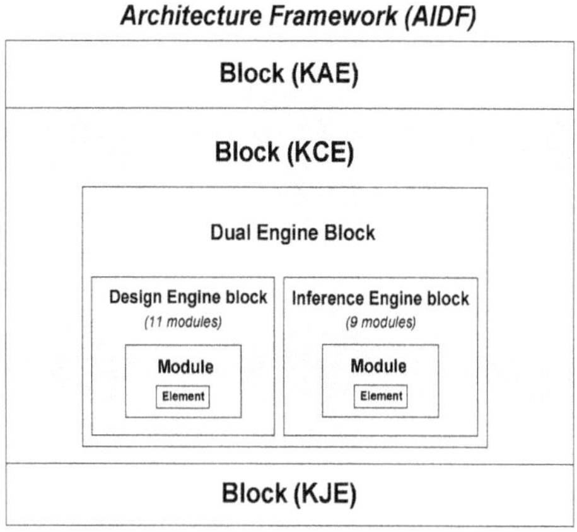

Architecture Framework (AIDF)

Block (KAE)

Block (KCE)

Dual Engine Block

Design Engine block (11 modules)

Inference Engine block (9 modules)

Module

Element

Module

Element

Block (KJE)

Fig. 3.17. AIDF divided into blocks, engines, modules, and elements.

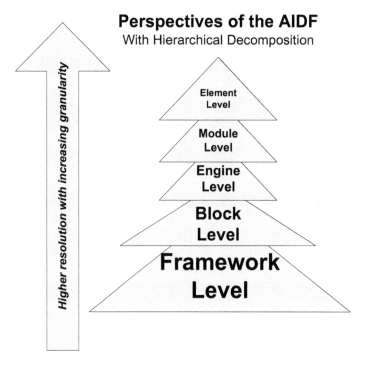

Perspectives of the AIDF
With Hierarchical Decomposition

Element Level

Module Level

Engine Level

Block Level

Framework Level

Higher resolution with increasing granularity

Fig. 3.18. Increasing granularity of the AIDF providing more perspective analysis.

BLOCK OPERATIONS MAPPED TO ENGINES

Input block Functioning as Knowledge Assimilation Engine (KAE)

The AIDF *input block* is involved in general web-based global domain knowledge acquisition using intelligent agents and proactive *a prior* knowledge entry from distributed experts in the field, as well as gathering current data using standard entry techniques on a particular project from a user, usually a systems engineer. The designers benefit from advances made globally from a proactive retrieval process which fetches knowledge units, as needed, from network sources around the world.

Processing Block Functioning as Knowledge Correlation Engine (KCE)

The AIDF *processing block* is involved in processing numeric, logical, and algorithmic engines, i.e. providing complexity and numerical analysis, as well as engineering synthesis, comprised of a validation engine, design engine, and AI engine involved in executing domain rules, algorithmic calculations, and component searches in the distributed knowledge base (KB) using SWS technology and ontology reasoners provided by technology, such as Jena [Reynolds, 2005]. These engines can be selectively activated and accessed with an MVC displaying the deductive/inductive reasoning capability using backward/forward chaining mechanisms, orchestrated by the validation modules ensuring the practice of validated industry standards, design modules ensuring the practice of sound design theory, and AI modules ensuring justifiable recommendations. The benefits of having a modular architecture in addition to uncoupling of the three main engines in the processing block creates a logical separation of concerns, providing capability for independent KB development by autonomous intelligent agents crawling the Web, as well as human domain experts proactively updating knowledge in their respective areas of validation standards, design theory, artificial intelligence algorithms, and particular case study knowledge.

Output block Functioning as Knowledge Justification Engine

The AIDF *output block* displays justifiable recommendations and engineering configurations after real-time processing. The output block in a previous stage becomes the input to the next stage, after passing through the intermediate gate.

Once the design process is complete, a proof-of-concept prototype can be implemented using Rapid Application Development and Dynamic Systems Development Method (DSDM) [Schulmeyer and MacKenzie , 1999], for instance.

BLOCK DESCRIPTIONS

Input Block (KAE)

The AIDF input block is responsible for gathering and storing design requirements from the systems engineer, while developing the knowledge base through domain expertise support and intelligent agent mechanisms. Each component's design risk weight is computed with respect to criteria such as cost, weight, dimensions, reliability, availability, maintainability, and dependability, factoring in optical component polarization, reflection, and diffraction losses. The effectiveness of an AI design environment for optical backplane engineering is largely determined by the quantity and quality of knowledge available *a prior* for machine processing, in addition to the speed of accessibility and deductive logic capability. To this end, the AIDF input block provides an interactive mechanism for large-scale maintenance and appropriate application of designer team rationale and project requirements using database technology infused with Semantic Web Service-enhanced KBE methodology that allows domain experts anywhere in the world to update the knowledge base through the Web.

Thus the power of the decision support available to the system designer grows in parallel with the expansion of the knowledge base, which grows organically through a combination of automated intelligent agent scouring the Web for requirements information, as well as periodic domain expert input from any point of presence. This enables both the intelligent agents and domain experts to systematically update the knowledge base (KB) with the most current system standards and authenticated ontologies, while locating the most appropriate OBIT component and layout design parameters available for incorporation into the central knowledge base. In order for all this collected knowledge to be processed, an algorithm block is needed.

Processing Block (KCE)

Once this relevant project information and *a prior* knowledge is accumulated, the designer needs a validated intelligent design environment that would be capable of automatically guiding the conceptual design process towards a logical conclusion by applying design risk mitigation techniques. This type of decision support is based on sound, established, and/or proven design theory, in addition to applying the aforementioned Semantic Web Services-enhanced KBE approach. To this end, the processing block does provide an effective approach to enhance system reliability by applying critical validation, design, and AI principles throughout the conceptual design process, via validation engine, design engine, and AI engine. In the case for FSOI, design risk mitigation at the conceptual design stage is provided by an automated mechanism for continuous verification to ascertain the level of design complexity by a combination of design theories and vulnerability analysis techniques, namely Axiomatic Design Theory (ADT), the Theory of Inventive Problem Solving (TRIZ), Design Structure Matrix (DSM), Fault Tree Analysis (FTA), Failure Mode and Effects Analysis (FMEA), and Conant entropy analysis based on *a prior* assessment of extensive component history stored in the KB inventory. Hence, the most reliable FSOI configuration can be recommended to the designer by intelligently applying theoretical axioms, methods, and techniques automatically.

Output Block (KJE)

The justifiable and traceable recommendations based on the inference engines [CLIPS, 2006] and calculations are provided in the output block. The detailed output can be provided by displaying the most reliable configuration producing the least complexity as represented by the entropy calculations provided by the Conant transmission graphs [Conant, 1972]. Also, the design analysis reflected by ADT, TRIZ, DSM, FTA, and FMEA can be used to determine the most reliable components or design parameters to fulfill the functional requirements.

AIDF ARCHITECTURAL FRAMEWORK MODEL DETAIL

ARCHITECTURE MODELS

AIDF architecture can be categorized into five models, namely *process, frame-work, structural, dynamic,* and *functional* model (Table 2). The emphasis of this thesis is on the AIDF architectural framework model. Although all models are not necessary to describe any given architecture, we leverage all models for maximum architectural description of the AIDF, independent of technology choice consistent with MDA approach: (1) The AIDF *process model* provides a methodology to develop the AIDF according to the 8-step axiomatic design V-Model. (2) The AIDF *framework model* provides an overall representation of the modular AIDF in terms of components and connectors. (3) The AIDF *structural model* provides various views describing the relationships and rationale for the object-oriented structural elements developed with standard UML 2.0 and extended with SysML 1.0, a new visual modeling language for systems engineering [UML, 2005]. (4) The AIDF *dynamic model* provides dynamic views describing the interactive behavior and dynamic evolution of the system, complementary to the framework and structural model. (5) The AIDF *functional model* provides functional views that can be represented directly with Acclaro DFSS framework software (Table 3.3).

Table 3.3. AIDF description represented by architecture framework model.

#	Architecture Model Type	Architectural Expression	Recommended AIDF Implementation method	Description
1	*Process Model*	Framework	8-step Axiomatic Design V-Model	Provides framework development methodology of architecture
2	*Framework Model*	Framework	Acclaro Design for Six Sigma (DFSS)/MS Visio	Provides an overall representation of architecture
3	*Structural Model*	ADL/OO diagrams	SysML 1.0/ UML 2.0 /MML /CommonKADS	Provides specific OO views expressing how each framework module is constructed using an ADL
4	*Dynamic Model*	ADL/OO diagrams	SysML 1.0/UML 2.0/MML SWS/Visual OWL/CmapTools/model-view-controller	Provides OO views expressing interactive and dynamic parts
5	*Functional Model*	Framework	Acclaro DFSS	Provides views expressing functionality

Hence, taken together, all five architectural models representing the AIDF advance the art of computer engineering by providing a powerful conceptual approach to decision support for an AI design environment demonstrated by the case study on optical backplane engineering. Thus, starting with the Axiomatic Design process V-model, all the other models are then developed using Acclaro DFSS and Visio, in conjunction with TeleLogic TAU Generation 2 [Telelogic, 2006]. Other advanced tools, modeling techniques, and global standards for expert systems, ontological development, and knowledge engineering were explored, such as Matlab Fuzzy Logic and Bayesian Toolbox, MML, CommonKADS, CmapTools [Concept Maps, 2006], JESS [Sandia National Laboratories, 2006], and Semantic Web Services Technology, such as Visual OWL and HP Labs [W3C, 2006].

AIDF ENGINES

The AIDF environment for automated intelligent system design has an architectural framework model comprised of three types of engines, namely KAE, KCE, and KJE (Fig. 3.4). During the design process, the framework model undergoes three stages of design, in which functional requirements and design parameters are handled in the conceptual design stage, layout and component integration considerations are handled in the embodiment design stage, and optimization choices are handled in the detailed design stage.

However, it is emphasized that throughout all stages, the *KAE, KCE,* and *KJE* are operational (Table 3.4), supporting the needs of each stage, depending on the stage, by managing all types of project and domain knowledge in the KAE, processing the data in the KCE, and displaying justifiable and traceable recommendations in the KJE (Table 3.3). This feature allows the design rational to be provided immediately as a form of intelligent feedback based on domain expert knowledge either retrieved from local or networked sources, with the added benefit of inference mechanisms initiated in the dual engine block.

Table 3.4. Description of the engines engaged in each stage.

Engine Type	Description (Purpose)
Knowledge Assimilation Engine (KAE)	To assimilate knowledge by collecting, gathering, and combining knowledge from user and environment
Knowledge Correlation Engine (KCE)	To correlate knowledge by coordinating, tasking, and processing the influx of knowledge from the KAE with the AI design engine block
Knowledge Justification Engine (KJE)	To justify the automated recommendations by providing an inter-action mechanism using an MVC, justifiable recommendations, and a rich visualization display

KAE Overview

The KAE is involved in managing both domain expert and user input and preparing it for processing in the KCE. Managing input from distributed domain experts in the field is accomplished through a web-based knowledge acquisition system that dispatches intelligent agents to proactively gather and update the knowledge base with *a prior* knowledge. The type of state-of-the-art technology recommended for achieving maximum knowledge acquisition is Semantic Web Services (SWS) that is capable of assisting the design engineer by recommending solutions provided by domain experts in the field via templates.

KCE Overview

The KCE is involved in processing the data transferred to the block from the KAE and preparing it for output in the KJE. Processing the data and knowledge is done using engines that can make critical engineering calculations in the design engine block and algorithmic calculations in the AI (inference) engine block. Automated engineering analysis and synthesis support is provided by a task manager and specific support modules, activated by the task manager depending on need. The design engine provides all types of design support for risk mitigation through complexity reduction and numeric analysis. The AI (inference) engine is involved in executing domain rules, algorithmic calculations, and component searches in the distributed (networked) knowledge base (KB) using SWS technology and ontology reasoners.

KJE Overview

The KJE is involved in justifying the data transferred to the block from the KCE by using a Model View Controller (MVC) architecture paradigm, allowing the user to selectively view, access, and interact with the model in operation. Justification of the design decisions and recommendations can be provided to the user by displaying the background reasoning process done to reach a particular conclusion. Some examples of artificial thought processing displayed could be, for instance, deductive/inductive reasoning and backward/forward chaining mechanisms. Further justification can be provided by displaying the mechanism used for validation in several ways. For instance, the system can demonstrate that validated industry standards were followed by referencing KB rules focusing on validation issues that were applied during the reasoning process, such as validated industrial best practices or design theory. Hence justifiable recommendations, engineering guidelines, and optimized configuration blueprints are provided by the KJE, at the end of the third stage. In summary, we show the detail of the AIDF here (Fig. 3.19).

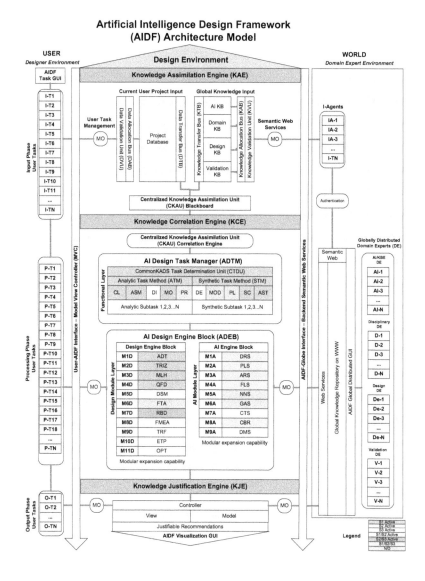

Fig. 3.19. AIDF architectural framework model detail.

AIDF KNOWLEDGE ASSIMILATION ENGINE

KAE OVERVIEW

The AIDF input block Knowledge Assimilation Engine (KAE) is responsible for gathering and storing design requirements from the systems engineer, while developing the knowledge base through domain expertise support and intelligent agent mechanisms shown in detail (Fig. 3.20) with components (Table 3.5). In addition to gathering functional requirements and design parameters, exact component specifications that are cu r-rently available can be gathered by the dispatched agents, providing a robust description of each component in terms of various parameters, such as cost, weight, dimensions, reli-ability, availability, maintainability, and dependability, as needed. The effectiveness of an AI design environment is largely determined by the quantity and quality of knowledge available *a prior* for machine processing, in addition to the speed of accessibility and d e-ductive logic capability. To this end, the AIDF input block provides an interactive mechanism for large-scale maintenance and appropriate application of designer team ra-tionale and project requirements using database technology infused with Semantic Web Service-enhanced KBE methodology that allows domain experts anywhere in the world to update the knowledge base through the Web via easy-to-follow and evolving templates that meet the needs of knowledge description.

Thus, the power of the decision support available to the system designer grows in parallel with the expansion of the knowledge base, which, in turn, grows organically through a combination of automated intelligent agents scouring the WWW to meet AIDF user requirements, as well as periodic domain expert input from any point of presence. This approach enables both the intelligent agents and domain experts a) to systematically update the knowledge base (KB) with the most current system standards and authenti-

cated ontologies, and b) locate the most appropriate component and layout design pa-

rameters available for incorporation into the central knowledge base.

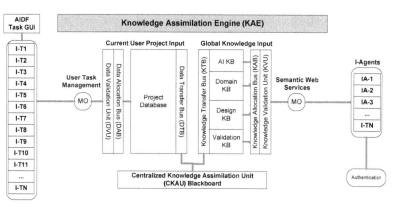

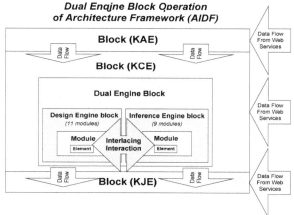

Fig. 3.20. Knowledge assimilation engine framework detail.

Table 3.5. Knowledge assimilation engine components.

Engine 1	Components
Knowledge Assimilation Engine (KAE)	Semantic Web Services Monitoring unit
Auxiliary Intelligent Agents (I-Agents)	Global Knowledge Input - Knowledge Validation Unit (KVU) - Knowledge Allocation Bus (KAB) - Knowledge Base
Authentication	- *AI KB* - *Domain KB*
Semantic Web Services	- *Design KB* - *Validation KB* - Knowledge Transfer Bus (KTB)
Web Services	User Task Management
Global Knowledge Repository on WWW	Monitoring unit
Globally Distributed Domain Experts (DE)	Current User Project Input - Data Validation Unit (DVU) - Data Allocation Bus (DAB) - Project Database - Data Transfer Bus (DTB)
- *AI/KBE DE* - *Disciplinary DE* - *Design DE* - *Validation DE*	Centralized Knowledge Assimilation Unit (CKAU) *Blackboard*
AIDF Globally Distributed GUI	

KAE DETAIL

The types of knowledge that is acquired by the Knowledge Assimilation Engine (KAE) are all types of knowledge that will be used to determine best design characteristics, e.g. functional requirements, design parameters, constraints, various design layout libraries, component simulation information, and optimization techniques. The functional requirements are provided by the designer based on the project needs through a graphical user interface to the AIDF based on a model-view-controller (MVC) paradigm.

As the needs are systematically provided as input, the AIDF automatically monitors the designer requirements and requests further elaboration, as the need arises, in order to clarify a requirement. Meanwhile, as the design parameters, layouts, components, etc. are selected by the designer, alternative options are supplied, as needed, through recommendation, by the knowledge base. This knowledge base is connected to the world via Web Services, where intelligent agents are dispatched to collect knowledge in the form of ontologies marked up in languages such as OWL that may meet the needs of the design. The knowledge is available in knowledge repositories on the Web and patent databases, which are sources of authenticated knowledge. The KAE can be partitioned into two function concerns (1) *Global Knowledge Input* and (2) *Current User Project Input.*

Global knowledge input (Table 3.6) begins with domain experts providing knowledge via the Web that finds its way to the AIDF through an intricate process. Long before current project data is provided by the designer, *a prior* knowledge is provided to the AIDF and continuously updated by the domain experts. Individual Globally Distributed Domain Experts (DE) in four main areas update this knowledge: (1) AI/KBE domain experts who are proficient in artificial intelligence matters, (2) Disciplinary domain experts who are proficient in any given discipline, (3) Design domain experts who are proficient in design matters, (4) Validation domain experts who are proficient in validation matters. The *AIDF Globally Distributed GUI* provides a convenient means for these domain experts to continuously update the AIDF with new knowledge from any point of presence, which is deposited in the *Global Knowledge Repository on WWW*. A new emerging technology, called *Semantic Web Services,* provides a means to encode all types of domain knowledge available in the form of ontologies into machine-processable format. Once this knowledge undergoes an *authentication* process, then the knowledge must be recognized and fetched by intelligent agents having semantic capability, i.e. *I-Agents.* There can be many types of I-Agents which are specialized for fetching certain types of knowledge, as designated by *I-A1, IA2, I-A3...I-AN*

All types of global knowledge and information is pushed and pulled to the AIDF system to be processed in each engine, KAE, KCE, and KJE, and subsequently displayed, through Web Services operating in real-time, throughout every operational moment of each stage. While this useful information is on display for review by the designer, the AIDF monitors the progression of the domain expert input around the world through a mechanism dedicated to exploiting *Semantic Web Services* in order to continuously update the AIDF knowledge base with new, authenticated, and validated information. Hence, the data is transferred by pipeline to the *Knowledge Validation Unit (KVU),* where the incoming data is validated. Once the data is processed in the KVU, the *Knowledge Allocation Bus (KAB)* takes control and allocates the knowledge into the appropriate memory buffers in the Knowledge Base. Four types of knowledge are maintained: (1) the *AI KB* contains knowledge dedicated to artificial intelligence, (2), the *Domain KB* contains knowledge dedicated to disciplinary domain expertise (3) the *Design KB* contains knowledge dedicated to design theory (4) the *Validation KB* contains knowledge dedicated to validation, in terms of the latest standards and metrics. A mechanism for supplying the designer with knowledge on demand, as needed, is provided by the *Knowledge Transfer Bus (WTB),* which sends this information to the *Centralized Knowledge Assimilation Unit (CKAU) Blackboard,* where the knowledge is combined with the data provided by the designer. The CKAU functions as a knowledge repository that prepares the

information collected from both the designer and world for processing in the Knowledge
Correlation Engine (KCE).

Table 3.6. Global knowledge input.

Global Knowledge Input	Purpose
Intelligent Agents (I-Agents)	Provides a means to fetch knowledge form WWW
Authentication	Provides a means to authenticate knowledge fetched by the I-Agent
Semantic Web Services	Provides a means to monitor and exploit Semantic Web Services
Knowledge Validation Unit (KVU)	Provides a mean to validate incoming global knowledge
Knowledge Allocation Bus (KAB)	Provides a means to allocate incoming knowledge into the KB
AI KB	Provides a means to store AI knowledge
Domain KB	Provides a means to store various domain knowledge
Design KB	Provides a means to store design knowledge
Validation KB	Provides a means to store validation knowledge
Knowledge Transfer Bus (KTB)	Provides a means to transfer knowledge to CKAU blackboard, as needed
Centralized Knowledge Assimilation Unit (CKAU) Blackboard	Provides a means for assimilation of knowledge coming from the designer, domain experts, and the WWW
Web Services	Provides a means to display pertinent information that can support the designer

Current User Project Input (Table 3.7) begins with the designer providing particular project data directly to the AIDF through an intricate interactive process, whereby the artificial intelligence operates in the background, monitoring all user activity. As the data is
provided by the designer via *User Task GUI* in the form of Input Tasks (I-T1, I-T2, I-

T3…), the AIDF monitors the progression of the input phase through *User Task Management.* Then the data is transferred by pipeline to the *Data Validation Unit (DVU),* where the incoming data is validated. Once the data is processed in the DVU, the *Data Allocation Bus (DAB)* takes control and allocates the data into the appropriate memory buffers in the *Project Database.* After the project database reaches critical mass, i.e. after sufficient data has been collected on the current project requirement specifications, the *Data Transfer Bus (DTB)* sends this information to the *Centralized Knowledge Assimilation Unit (CKAU) Blackboard,* where the knowledge is combined with the knowledge from the domain experts and the WWW.

Table 3.7. Current user project input.

Current User Project Input	Purpose
AIDF Task GUI	Provides a mechanism to interact with Designer
User Task Management	Provides a mechanism to monitor designer activity
Data Validation Unit (DVU)	Provides a mechanism to validate designer input
Data Allocation Bus (DAB)	Provides a mechanism to allocate data into project database
Centralized Knowledge Assimilation Unit (CKAU) Blackboard	Provides a mechanism to assimilate designer input

AIDF KNOWLEDGE CORRELATION ENGINE

KCE OVERVIEW

Engine Block Detail with Modules

Processing begins in the Knowledge Correlation Engine (KCE), whose details are shown (Fig. 3.21) with components (Table 3.8). Once this relevant project information and *a prior* knowledge are accumulated by the KAE, the designer needs an intelligent design environment that would be capable of monitoring his tasks during this process. This capability offers further analytic or synthetic decision support as needed, which automatically guides the design process towards a logical conclusion, while providing risk mitigation strategies, tactics, and techniques.

The KCE is involved in correlating the knowledge already assimilated by the AIDF in the KAE. This entails the appropriate assignment of a task based on the CommonKADS knowledge modeling approach in the CKAU Correlation Engine. This is followed by an appropriate selection of a subtask, which, in turn, selects and executes the appropriate set of configured modules in the ADEB.   More on configuration will be provided later.

Validation of this type of decision support that emphasizes system reliability is provided by sound, established, and/or proven design theory and state-of-the-art CommonKADS approach to knowledge-based engineering, which is a form of expert system in AI specialized for engineering needs.   Design risk mitigation is provided by the AIDF through a combination of design theories, vulnerability analysis techniques, and algorithms provided by the design engine block and AI engine block containing any number of executable modules, based on established standards and methodologies.

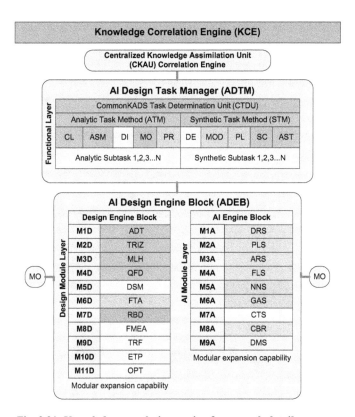

Fig. 3.21. Knowledge correlation engine framework detail.

Dual Engine Block Operation

The dual engine block (Fig. 3.21) can operate in tandem after connecting, or interlacing, the modules together (Fig 3.22). Later, we will show how to interlace the engine block for configuring the OBIT application for intra-module and inter-module operations within the same engine block and between the engine blocks using the framework. Furthermore, we will show how to interlace the elements, specified by the configured AIDF, for interaction within modules, as well as interaction between modules, for specifying intra-module and inter-module operation, respectively. These interactions represented by the framework-level, engine-level, module-level, and element-level can later be used to develop UML that meets the highly specified framework's requirements, as in the case specified for reliability engineering in the engine block. The interlacing interaction in the dual engine block provides a methodology to systematically integrate all the modules into a coherent pattern for problem solving. Thus, as one primary module becomes active, it may also, in turn, activate other modules as a way to provide auxiliary support to the inference process for more robust decision-making. Although more than one recommendation for the same scenario could be given, the strength of this approach provides the designer a wider selection of options, including design rationale and trade-off analysis, as would be encountered in a board meeting with differing domain expert opinions.

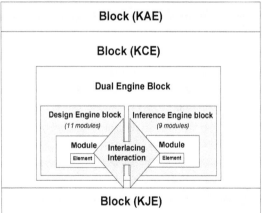

Fig. 3.22. Dual engine operation of AIDF.

Table 3.8. Knowledge correlation engine components.

Engine 2	Components
Knowledge Correlation Engine (KCE)	- Centralized Knowledge Assimilation Unit (CKAU) *Correlation Engine* - AI Design Task Manager - CommonKADS Task Determination Unit (CTDU)
Auxiliary Web Services	<u>Analytic Task Method (ATM)</u> - *Classification* - *Assessment* - *Diagnosis* - *Monitoring* - *Prediction* <u>Synthetic Task Method (STM)</u> - *Design* - *Modeling* - *Planning* - *Scheduling* - *Assignment* - AI Design Engine Block (ADEB) - <u>Design Engine Block Modules</u> - Axiomatic Design Theory (ADT) - Theory of Inventive Problem Solving (TRIZ) - Multi-Layered Hierarchy (MLH) - Quality Function Deployment (QFD) - Design Structure Matrix (DSM) - Fault Tree Analysis (FTA) - Reliability Block Diagram (RBD) - Failure Mode Effects Analysis (FMEA) - Technology Risk Factor (TRF) - Entropy (ETP) - Optimization (OPT) - <u>AI Engine Block Modules</u> - Domain Rule Support (DRS) - Predicate Logic Support (PLS) - Algorithm Reasoning Support (ARS) - Fuzzy Logic Support (FLS) - Neural Network Support (NNS) - Genetic Algorithm Support (GAS) - Conant Transmission Support (CTS)

KCE DETAIL

The *Centralized Knowledge Assimilation Unit (CKAU) Correlation Engine* is involved in correlating the knowledge sent to it from the CKAU blackboard. Once the KAE has managed to prepare the knowledge for processing, the Knowledge Correlation Engine (KCE) begins to correlate the knowledge assembled in the blackboard. This process is essential for matching the knowledge with the appropriate task executed by the module.

Based on CommonKADS approach to KBE, a knowledge model has three parts, each capturing a related group of knowledge structures, called a *knowledge category*, namely Domain Knowledge, Inference Knowledge, and Task Knowledge (Table 3.9). The Domain Knowledge is acquired in the KAE and KCE by monitoring the DE activity through Web Services. The Inference Knowledge, also acquired in the KAE and KCE, provides a way for the DE to supply the AIDF with executable rules to act on the domain knowledge. The Task Knowledge is supplied to the KCE.

The AIDF *AI Task Manager* provides a framework for analysis and synthesis tasks that is further subdivided into more subtasks, i.e. *synthesis/design* can have a *configuration* subtask (Table 3.10). As the number of functions increase in the AIDF, the type of tasks handled by the system can be increased. This allows the flexibility for the system to handle new tasks as the need arises.

Table 3.9. CommonKADS knowledge category definitions.

#	Knowledge Categories	Definition
1	Domain Knowledge	Composed of domain types, rules, and fact that specify the domain-specific knowledge and information that we want to talk define in an application, including relationships between knowledge types (i.e. ontologies); sometimes called a data model or object model
2	Inference Knowledge	Composed of basic inference mechanisms and roles that describe the basic inference steps using the domain knowledge. Inferences are best seen as the building blocks of the reasoning machine. In software engineering terms, the inferences represent the lowest level of functional decomposition.
3	Task Knowledge	Composed of task goals, decomposition, and control mechanisms that describe what goal(s) an application pursues, and how these goals can be realized through a decomposition into subtasks and (ultimately) inferences. Task knowledge is similar to the higher levels of functional decomposition in software engineering.

Table 3.10. CommonKADS analysis and synthesis definitions.

Type of Knowledge Intensive Task	Description	SubTasks Breakdown
Analysis	A type of super task requiring breakdown of information into constituent parts, enabling a focused investigation	Classification Assessment Diagnosis Monitoring Prediction
Synthesis	A type of super task requiring combination of parts or elements so as to form a whole, enabling a general symbiosis of parts	Design/configuration Modeling Planning Scheduling Assignment

Knowledge intensive tasks include *analytic* tasks and *synthetic* tasks (Fig. 3.23). The knowledge intensive tasks is comprised of analytic and synthetic tasks, which have appropriate sub-classifications, based on type of reasoning employed using the Common-KADS approach to KBE.

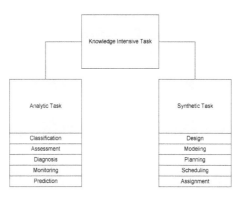

Fig. 3.23. CommonKADS standard modeling task decomposition.

There are five types of *analytic* tasks, namely classification, diagnosis, assessment, monitoring, and prediction (Table 3.11) [Schreiber et al. , 1999]. *Classification* is defined as an analysis task involving characterization of an object in terms of class. *Diagnosis* is defined as an analysis task involving determination of faults in the system. *Assessment* is defined as an analysis task involving characterization of a case in terms of a decision class. *Monitoring* is defined as an analysis task involving analysis of a system of a dynamic nature. *Prediction* is defined as an analysis task involving analysis of current system behavior to construct a description of the system state at some future point in time. Although there are many more possible types of analysis tasks for the AIDF, CommonKADS provides a foundation to start a categorization process.

Table 3.11. CommonKADS analysis task types.

Analysis	Description
Classification	An analysis task involving characterization of an object in terms of class
Diagnosis	An analysis task involving determination of faults in the system
Assessment	An analysis task involving characterization of a case in terms of a decision class
Monitoring	An analysis task involving analysis of a system of a dynamic nature
Prediction	An analysis task involving analysis of current system behavior to construct a description of the system state at some future point in time.

There are five types of *synthetic* tasks, namely design, modeling, planning, scheduling, and assignment (Table 3.12) [Schreiber et al., 1999]. An example of a subtask of *design* is *configuration*. *Design* is defined as a synthetic task such that the system to be constructed is a physical artifact. *Assignment* is defined as a synthetic task such that a mapping between a set of objects is made. *Planning* is defined as a synthetic task such that physical object construction is done having time dependencies, giving a sequence of activities. *Scheduling* is defined as a synthetic task such that planned activities are allocated resources. *Modeling* is defined as a synthetic task such that an abstract description of a system is constructed. Although there are many more possible types of synthesis tasks for the AIDF, CommonKADS provides a foundation to start a categorization process.

Table 3.12 CommonKADS synthesis task types

Synthesis	Description
Design	A synthetic task such that the system to be constructed is a physical artifact
Assignment	A synthetic task such that a mapping between a set of objects is made
Planning	A synthetic task such that physical object construction is done having time dependencies, giving a sequence of activities
Scheduling	A synthetic task such that planned activities are allocated resources
Modeling	A synthetic task such that an abstract description of a system is constructed

Once task determination is complete in the task manager, and an appropriate task is selected and activated by the AIDF, the *master control*, the primary system power allocation of the system, shifts to the *AI Design Engine Block (AIDEB)*. In AIDEB the modules are housed and the appropriate module is automatically selected for activation from either the *AI Engine,* housing the AI modules (Table 3.14), or *Design Engine,* housing the design modules (Table 3.13). Once the module is activated to meet the needs of the *control task*, i.e. the task currently in command of the operation of the module, algorithmic calculations ensue according to the respective rules and inference mechanisms encoded in the structural model (e.g. in SysML).

User task monitoring and domain expert knowledge monitoring occur throughout the design process. As the AIDF operates in the background, *monitoring* of the user activity (systems engineer) is continuously in operation in order to continue the design process without interruption, while the results of that particular task calculations are sent to the Knowledge Justification Engine (KJE) for feedback display. Continuous *monitor-*

ing for new knowledge, either using *Web Services* for general global knowledge retrieval, or specifically *Semantic Web Services* for domain expert knowledge retrieval, will be employed to gather newly updated world knowledge. However, when the modules are in execution, a system *lockdown* is initiated, whereby the parallel KB updating mechanism will not affect the current processing of the knowledge and consequent recommendations. However, if it turns out that the newly updated knowledge may affect the conclusions of the previous chain of logic, then the module can be activated again, e.g. the PLS and DRS module in the AI engine block could apply deductive and inductive reasoning again to derive an alternative conclusion with justification.

Another scenario could occur in the case of a faulty design detected by the AIDF in the early stages of the design process. For instance, if vital parts of the system being designed are moving towards an intolerably higher state of entropy, i. e. higher state of faulty disorganization, then color-coded warning lights will be displayed in the visual display, which is provided in the KJE (explained in the next section). These lights will provide feedback to the designer on the state of the system being designed in the form of a variable metric that would provide advance warning for risk mitigation purposes. The KCE functions as the processor of the data to detect the faults or system anomalies, ranging from minor glitch to catastrophic failure, allowing the AIDF to detect the potential malady and provide justification for the pinpointed weakness in the system. For instance, if the *diagnosis* task is activated, which executes the FTA and the FMEA module, then a Boolean fault determination algorithm will be initiated by the FTA module to determine the precise location of the fault, followed by the FMEA reporting on the most likely reason of failure. These type of raw calculations completed in the KCE are provided as results to the KJE *asynchronously* to preserve real-time operation of the feedback mechanism. Hence, this architecture framework allows for maximum capacity of parallel processing for all the tasks in the task manager that can be activated based on user actions, where the AIDF system orchestrates master control of the task manager, as needed, to the executable modules, which actually carry out the commands automatically in the background, throughout the design process. In other words, processing and feedback can happen in parallel at the same time for different tasks, as each task is executed and completed, and prepared for visual display.

Table 3.13. AIDF design engine module functions.

Design Modules	Acronym	Module Function
ADT	Axiomatic Design Theory	Provides an automated mechanism for hierarchical decomposition of FR and DP, provides 2 axioms, 11 corollaries, and 23 theorems for the rules base stored in the AI Engine Block
TRIZ	Theory of Inventive Problem Solving	Provides an automated mechanism for invention, especially by searching the Semantic Web for appropriate DPs
MLH	Hierarchical Multi-layer [design]	Providing an automated mechanism for going from FR to DP to components, calculation of reliability and cost
QFD	Quality Function Deployment	Provides an automated mechanism to ensure the customer guidelines are included in the quality of the design
DSM	Design Structure Matrix	Provides an automated mechanism to determine component to component interaction
FTA	Fault Tree Analysis	Provides an automated mechanism to predict component failures, where the calculations are based heavily on quantitative Boolean operators
RBD	Reliability Block Diagram	Provides an automated mechanism to estimate system reliability, where the calculations are based heavily on various engineering equations, such as MTTF (Mean time to Failure)
FMEA	Failure Mode Effects Analysis	Provides an automated mechanism to associate weights for each type of failure to assess fault qualitatively and trace root cause
TRF	Technology Risk Factor	Provides an automated mechanism to assess individual component risk on a cluster, mainly by associating any given component with a multiplier that affects the DSM
ETP	Entropy	Provides an automated mechanism to assess level of disorganization in system design during design process
OPT	Optical Backplane Engineering Domain	Provides an automated mechanism to manipulate domain-specific knowledge for inference, specifically in field of optical backplane engineering

Table 3.14 AIDF inference engine module functions

AI Modules	Acronym	Module Function
DRS	Domain Rule Support	Provides an automated mechanism for domain rule support, in terms of executable rules used by the inference engine
PLS	Predicate Logic Support	Provides an automated mechanism for logic support
ARS	Algorithmic Reasoning Support	Provides an automated mechanism for miscellaneous algorithmic reasoning support
FLS	Fuzzy Logic Support	Provides an automated mechanism for fuzzy logic support
NNS	Neural Network Support	Provides an automated mechanism for neural network support
GAS	Genetic Algorithm Support	Provides an automated mechanism for genetic algorithm support
CTS	Conant Transmission Support	Provides an automated mechanism for component transmission support
CBS	Calibrated Bayesian Support	Provides an automated mechanism for calibrated Bayesian support
DMS	Data Mining Support	Provides an automated mechanism for data mining support

AIDF KNOWLEDGE JUSTIFICATION ENGINE

KJE OVERVIEW

Once the calculations for the KCE are made in a particular module, they are transferred to the KJE for display, details shown (Fig. 3.24) with components (Table 3.15). Hence, the rationale for the design decisions and recommendations are provided, along with access to a traceable chain of logic. The paradigm suggested for the AIDF is the *Model-View-Controller* (MVC) approach.

KJE DETAIL

The *Model* is the aspect of the AIDF responsible for providing the user a <u>structure</u> demarcating the framework model. The *View* is the aspect of the AIDF responsible for providing the user an observation-post allowing a view of the operations of the AI design system. The *Controller* is the aspect of the AIDF responsible for providing the user an interactive-medium allowing a mechanism to administer the system. Once the system modules in the KCE give results, the KJE will display the results with *Justifiable Recommendations* on the design. Optionally, an *AIDF Visualization GUI* can be added to show the results in a way that can be quickly visualized for expedient comprehension by the user throughout the design process. *Monitoring* of the user input data, of the domain experts via Semantic Web Services, and of the miscellaneous developments in the world are done by Web Services.

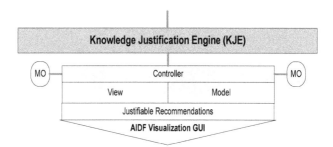

Fig. 3.24 Knowledge justification engine framework detail.

Table 3.15. Knowledge justification engine components.

Engine 3	Components
Knowledge Justification Engine (KJE)	- Model - View - Controller - Justifiable Recommendations
Auxiliary Web Services	- AIDF Visual Display

AIDF AUTOMATED DESIGN PROCESS SUPPORT OVERVIEW

DESIGN PROCESS STAGES

Design Phases Supported by Corresponding AIDF Stage

It has been generally accepted that the engineering process can be divided into three major design *phases* [Pahl and Beitz, 1988]. In the Artificial Intelligence Design Framework (AIDF), each of these design phases is supported by a corresponding AIDF *stage* defined as the *Conceptual Design Stage, Embodiment Design Stage,* and the *Detailed Design Stage* (Fig. 3.25). Thus, the AIDF architectural framework provides a structure to develop a KBE system based on a compartmentalized design process, attacking each stage separately and then passing on the resulting recommendations to the next stage.

For each stage in the AIDF design process, three engines are actively engaged in providing the necessary support associated with the stage. The goal of the first AIDF stage (CDS) is to assist the designer in making *qualitative* design decisions, generally regarding functional requirement and design parameter assessment. The goal of the second AIDF stage (EDS) is to assist the designer in making *quantitative* design decisions, generally regarding component layout and component considerations. The goal of the third AIDF stage (DDS) is to assist the designer in making both *quantitative* and *qualitative* design decisions, generally regarding optimization of the selected components.

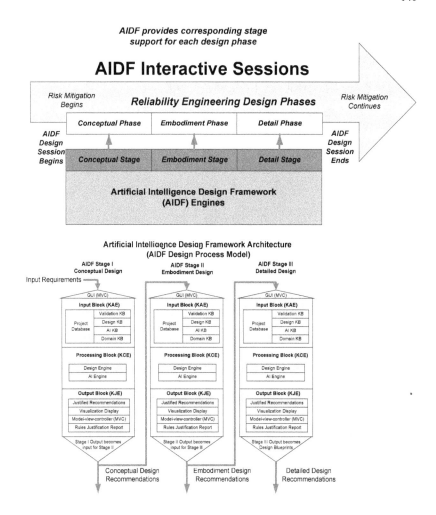

Fig. 3.25. AIDF provides corresponding design stage support for each design phase.

AIDF Provides Support in All Phases

Although greater impact (Fig. 3.26) on the final product is made in early design

phases, there are more design tools available supporting later design phases [Wang et al.,

2001]. However, KBE SoS applications developed by the AIDF are intended to provide full support in all three phases. The degree of impact of decisions made early on are critical to ensure that what is developed in later phases of the design process will eventually meet the needs of the customer, cutting down on cost overruns.

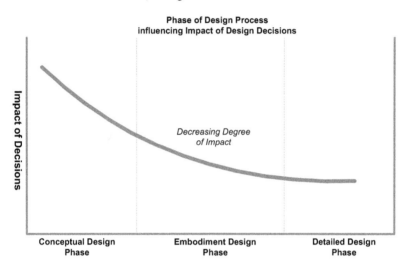

Fig. 3.26. AIDF supports all phases where degree of impact is greatest early on.

Stage Functions

The AIDF model addresses all major engineering design stages (Table 3.16) by adhering to a three-stage architecture framework defined by the engineering concerns of (a) *Conceptual Design* where the qualitative functional requirements and design parameters are decided, (b) *Embodiment Design* where quantitative structural and component issues are determined, and (c) *Detailed Design* where the final design blueprints, recommendations, and configuration are provided.

Table 3.16. AIDF conceptual design stage active engines.

AIDF Design Stage	AIDF Design Stage Definition	Active Engines
Stage I *Conceptual Design (CDS)*	Stage supporting the *selection* of fundamental design choices and methodology, predominantly in terms of *qualitative* functional requirements and design parameters	All three engine blocks: KAE, KCE, KJE
Stage II *Embodiment Design (EDS)*	Stage supporting the *development* of structural layout and component issues, predominantly in terms of *quantitative* analysis and assessment	All three engine blocks: KAE, KCE, KJE
Stage III *Detailed Design (DDS)*	Stage supporting optimized *integration* in terms of both *qualitative and quantitative* design blueprints, recommendations, and specific configuration guidance	All three engine blocks: KAE, KCE, KJE

Common Activities for Each Stage

All of three of the stages (Fig. 3.27) contain three blocks with separate but interrelated functions. The Knowledge Assimilation Engine (KAE) functions as an input block for current project data and long-term knowledge warehoused in the knowledge repositories, categorized and updated according to field, i.e. validation, AI, design, domain. Each field can be updated by thousands of domain experts in the field, with just as many intelligent agents operating between the AIDF and the knowledge repositories, accessing the knowledge asynchronously, as needed, during the design process. The project data is provided synchronously according to the stage of the design process. Both processes can work concurrently. The processing block function as the Knowledge Correlation Engine (KCE), integrating all forms of rules and ontologies in the blackboard for any given design project so that the engines may operate on them, either with brute-force design calculations or inference techniques. After processing, both of these engines produce recommendations that are submitted to the output block for display. The processing block contains a design engine and an inference engine which houses twenty modules specialized for design support, concentrating on reliability engineering. Each of these modules are connected to the networked knowledge repositories and accessed with intelligent agents. The output block provides justified recommendations, if needed, to the designer, based on the recorded design rationale. The visualization display provides various views of the recommendations and the rationale, such as design structure matrix module showing the component interactions and color-coded risk areas determined by the Conant transmission module. The model-view-controller provides a way for the design model and its associated reliability assessments to be displayed and interacted on by the designer. A rules justification report can be provided on demand that displays all the rules executed and their reasons.

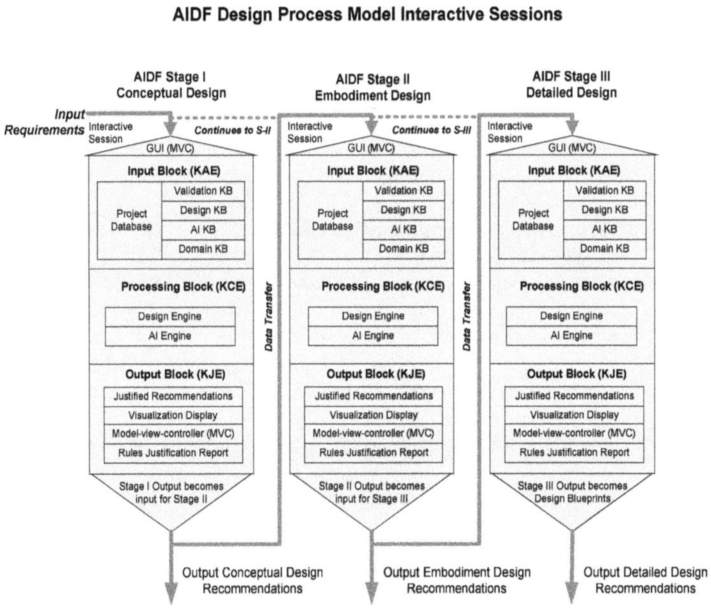

AIDF Design Process Model Interactive Sessions

Fig. 3.27. Design phases matching corresponding stages of AIDF.

INTERACTIVE SESSIONS ACTIVE FOR EACH STAGE

The AIDF engines provide design support in the form of risk mitigation for reliability engineering in three stages. Interactive sessions are provided for the design engineer, as well as the domain experts, to get assistance and provide domain expertise, respectively. Many of these processes are done concurrently, i.e. while the product designers are actively involved with a particular project utilizing a KBE SoS, domain experts can be proactively developing and updating the knowledge repositories (Fig. 3.28).

These processes can co-exist simultaneously, with intelligent agents providing synchronous design support with respect to the phases, while knowledge is asynchronously updated independent of any given design project. Knowledge engineering can be accomplished without the intervention of knowledge engineering using CommonKADS templates combined with Web technology using CmapTools as an ontology editor [Schreiber et al., 1999] [CmapTools, 2006]. We tested the concept of this type of ontology editor, whose screenshots are provided later.

Intelligent agents can be used to retrieve component models for a simulation during early parts of the design phase before actual production using a tool for reasoning with the Semantic Web [Kopena, 2003]. All of these processes are essential for a fully functioning KBE SoS application that fully leverages the knowledge available online, especially in knowledge repositories continuously being updated and authenticated by domain experts in advance. Thus, intelligent agents can be used to retrieve components marked up in OWL, for instance, for product engineering simulation as needed based on machine processible semantics, as defined by the requirements of any given project.

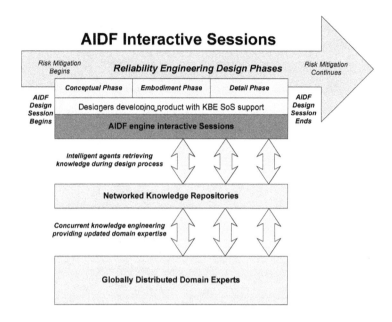

Fig. 3.28. Interactive AIDF sessions with KB network and agents.

ORDER OF PROCESSING FOR EACH STAGE

Compartmentalized Processing Sequence of Engines within Each Stage

The order of processing begins at the conceptual design phase in which the informal product requirements are elicited and translated into more formal functional requirements. During this phase, the AIDF CDS provides real-time intelligent support, primarily by assessing system requirements and providing alternative design choices, in the form of Design Parameter (DP) recommendations to resolve design complexity issues by decoupling and trade-off analysis. During this phase, the three blocks are engaged in this sequence: (KAE), (KCE), and finally (KJE) (Fig. 3.29). Once these blocks are complete for the first stage and management validates the output, the next stage may begin.

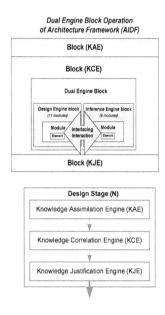

Fig. 3.29. AIDF design stage engines in sequence.

Compartmentalized Processing Sequence between Stages

The AIDF model is based on architectural standards terminology of ANSI/IEEE Std. 1471-2000 and model-driven architecture (MDA) concepts to ensure interoperability and stability at the highest levels of abstraction [OMG, 2006]. ANSI/IEEE Std. 1471-2000 provides interoperability by promoting terminological consistency and the MDA conceptual approach provides for architectural stability by emphasizing system functional requirements over implementation details by clear separation of concerns. The AIDF design process shows how the AIDF model supports the compartmentalization of the engineering design process by separation into three distinct design stages, where each stage is further subdivided unto functional blocks (Fig. 3.30).

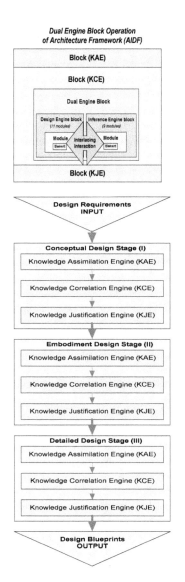

Fig. 3.30. AIDF timing sequence diagram.

AUTOMATED DESIGN PROCESS INTEGRATED WITH MANAGEMENT

RELIABILITY ENGINEERING IN ALL THREE STAGES

The AIDF supports all major stages in an engineering design lifecycle, where validated design theories and AI methodologies are continuously and automatically updated into its knowledge base by intelligent agents, in addition to proactive development by knowledge engineers and domain experts. The AIDF also provides intelligent assessment in early phases of conceptual design, where it is widely accepted by industry that 75% of cost is incurred [Wang et al., 2001]. Thus, the AIDF enables analytic decision support to provide design risk mitigation in the case study. This is accomplished by reliably resolving the high-speed communications bottleneck challenging optical backplane engineers with real-time decision support on OBIT concerns [Grimes, 2005]. This design risk mitigation is accomplished by numerical analysis, deductive and inductive inference techniques, and justifiable recommendations processed by the design engine and inference engine for reliable product engineering component optimizing on configuration and structural design. The intelligent feedback is based on validated engineering theory and algorithms updated by a globally distributed KB, immediately accessible by intelligent agents.

Validation at Gates

All stages have internal validation and feasibility processes, usually predefined by the systems engineer. If the resulting design recommendations did not pass the internal validation process, the first stage (CDS) is considered incomplete and redone until completed properly (Fig. 3.31). If the design is deemed infeasible at any time, the AIDF system will recommend an engineering project *exception handling* in which management involvement would be required and may lead to an *abort*. Upon validation, the *qualitative* design recommendations are transferred to the next stage, and a similar three-block engine process continues into the next stage (EDS), by taking the qualitative data as a starting point before initiating a *quantitative* analysis that predominantly provides analytic recommendations on design layout and choice of components. Then, the final design stage (DDS) is initiated which provides both a *qualitative* and *quantitative* analysis that results in recommendations based on state-of-the-art best practices in order to optimize the preceding layout and component selections. Hence, although each stage engages the same engine, KAE, KCE, and KJE, different types of design support and AI tasks are activated, based on the engineering phase. Once the AIDF design engines process all input requirements based on the appropriate knowledge sources, final design blueprints advising on reliability, are provided.

In addition to showing how the AIDF model supports engineering product design segregated into functional blocks, other peripheral operations such as graphical user interface interactions, validation and verification checks, engineering and management gates, and data transfer points are also provided for reference (Table 3.17). The three critical design stages encountered in product engineering are addressed with respect to management with the design process model showing the gates between the stages, namely *conceptual design, embodiment design, and detailed design*.

150

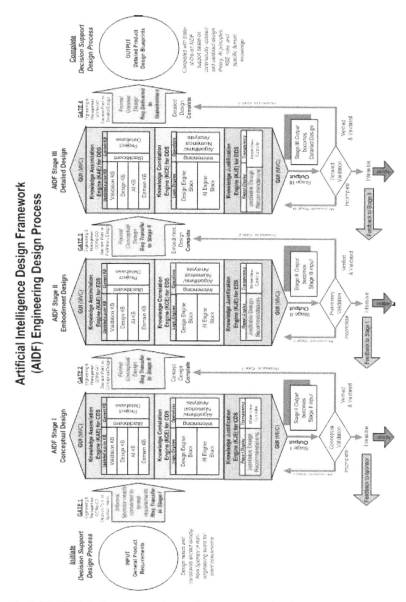

Fig. 3.31. AIDF design process stages with management gates.

Table 3.17. Major AIDF functional blocks and operations in each stage.

#	Blocks	Function	Mapping to AIDF Modular Functionality
1	*GUI*	Graphical User Interface mechanism	Provides a method for user interaction using MVC
2	*Input Block (KAE)*	Involved in keeping track of project and domain knowledge, i.e. general web-based global domain knowledge acquisition using intelligent agents and proactive *a prior* data entry from distributed experts in the field and particular project management interfacing with the user, usually a systems engineer.	Project Database maintains designer's project real-time knowledge Knowledge Base Type providing *scalable modules* that house 4 different types of knowledge: 1. Validation KB – houses validation knowledge 2. Design KB – houses design knowledge 3. AI KB – houses artificial intelligence, KBE rules, 4. Domain KB – houses case study knowledge (e.g. optical backplane engineering domain rules)
3	*Processing Block (KCE)*	Involved in processing numeric, logical, and algorithmic engines, i.e. providing complexity and numerical analysis, as well as engineering synthesis, comprised of a validation engine, design engine, and AI engine involved in executing domain rules, algorithmic calculations, and component searches in the distributed knowledge base (KB) using SWS technology and ontology reasoners.	Design Engine Processes logic governing current design theory, theorems, principles, risk mitigation algorithms, and techniques, and numeric analysis AI (inference) Engine Processes logic governing inference engines using deductive/inductive logic, algorithms, rule-bases, heuristics, KBE CommonKADS standards, case-based reasoning, neural network analysis, predicate logic, fuzzy logic, intelligent agent dispatching
4	*Output Block (KJE)*	Involved in displaying justifiable recommendations and engineering configurations after processing	Provides justifiable recommendations for design blue-prints, optimal component configuration, structural layout guidance, manufacturing suggestions, requirements tracing, case-based design history reports with intelligent guidance records
5	*V/V Checks*	Involves testing for verification and validation by engineers to see if design met requirements and real-world needs	Provides a method for design verification and validation by engineers and management after each design stage completion.
6	*Gates/Data Transfer*	Involves an assessment of stage design by management before approving data transfer	Provides a mechanism to account for political and cost concerns, in addition to engineering feasibility

CONCEPTUAL DESIGN STAGE RELIABILITY AUTOMATION

Conceptual Design Stage (CDS) supports the *selection* of fundamental design choices and methodology, predominantly in terms of *qualitative* functional requirements and design parameters. We will discuss the reliability automation of the engine block modules during CDS. An overview of the functions (Table 3.18) and active engines (Table 3.19) in conceptual design is provided.

ACTIVE RELIABILITY SUPPORT MODULES IN CDS BLOCK ENGINES

The three block engines in the conceptual design phase can be divided into functional blocks, namely Knowledge Assimilation Engine (KAE), Knowledge Correlation Engine (KCE), and Knowledge Justification Engine (KJE).

CDS/KAE Block 1 Function

The conceptual design stage (CDS) Knowledge Assimilation (KAE) is responsible for assimilating *qualitative* knowledge by collecting, gathering, and combining knowledge from user and environment. Design Parameters (DP) are retrieved and assembled from the global knowledge depository available on the WWW. In the first block during the CDS, the processing function of the KAE is one of *assimilation*, which entails gathering of conceptual elements of design such as Functional Requirements (FR), Design Parameters (DP), and Constraints in a way conducive to easy incorporation into the engineering design. In order to achieve this end, the active region of the AIDF system is the monitoring elements that scan for new developments by system users and world activity that could affect or improve the design process.

CDS/KCE Block 2 Function

In the second block during the CDS, the processing function of the KCE is one of *correlation*, which entails the matching of the needs of the designer to specific FR so that appropriate DPs can be selected by the activated CommonKADS KBE task manager that executes the appropriate analytic or synthetic task associated with a particular module, according to the constraints, by the support engines. The conceptual design stage is primarily concerned with analytic tasks, which invoke associated *Design Engine* modules such as the ADT, MLH, TRIZ, and QFD and corresponding *AI engine* modules such as DRS and FLS. More on the operations of these modules in reliability engineering based on the AIDF is provided later. The CDS Knowledge Correlation Engine (KCE) is responsible for correlating *qualitative* knowledge by coordinating, tasking, and processing the influx of knowledge from the KAE with the AI design engine block. In this block, DPs that are assembled by the CKAU *blackboard* are matched to appropriate tasks in the ADTM by the CKAU *correlation engine*. The ADTM is then responsible for coordinating which module is executed in the ADEB to meet the requirements of any given executable task in operation.

CDS/KJE Block 3 Function

In the third block during the CDS, the processing function of the KCE is one of *justification*, which entails display of the logical sequence of artificial thought patterns conducted by the AIDF system, such as deduction or induction, that arrive at any given conclusion to fulfill the FRs within constraints. The CDS Knowledge Justification Engine (KJE) is responsible for justifying the automated *qualitative* recommendations by providing an interaction mechanism using an MVC, justifiable recommendations, and a rich visualization display. In this block, DPs that are selected for use in the design, based on the current specifications (FR) that are provided as output, complete with full justification and rationale, which can be displayed using a windows environment. Hence, a series of good justifications by the AIDF system is expected to boost the confidence of the designer on a particular design recommendation, thereby swaying his decision in the right direction – all the while reducing risk through intricate risk mitigation calculations conveniently provided in the background. A *model-view-controller* paradigm can be used to achieve this type and level of interaction with the user, where the *model* provides a systematic basis for the operations, the view provides transparency to the operations of the model, and the *controller* provides a means to interact with the model. A sophisticated graphical visualization display can be connected to the output of the MVC, based on the level of current technology available.

Table 3.18. AIDF conceptual design stage overview.

CDS Engines	Purpose	Implementation schema
CDS/KAE (S1/Block 1)	To assimilate *qualitative* knowledge by collecting, gathering, and combining knowledge from user and environment	Design Parameters (DP) are retrieved and assembled from the global knowledge depository available on the WWW
CDS/KCE (S1/Block 2)	To correlate *qualitative* knowledge by coordinating, tasking, and processing the influx of knowledge from the KAE with the AI design engine block	DPs that are assembled by the CKAU *blackboard* are matched to appropriate tasks in the ADTM by the CKAU *correlation engine.*
CDS/KJE (S1/Block 3)	To justify the automated *qualitative* recommendations by providing an interaction mechanism using an MVC, justifiable recommendations, and a rich visualization display	DPs that are selected for use in the design, based on the current specifications (FR) that are provided as output, complete with full justification and rationale

Table 3.19. Active engines in AIDF/CDS.

Stage/Engine	Processing Function	Active Region
CDS/KAE (S1/Block 1)	*Assimilation* of conceptual elements of design such as Functional Requirements (FR), Design Parameters (DP), and Constraints	Monitoring of User Activity Monitoring of World Activity
CDS/KCE (S1/Block 2)	*Correlation* of FRs, DPs, and constraints with appropriate support engines in a way so that the designer's activity is supported by CommonKADS framework for KBE	Primary Active Task: Analytic Tasks Primary Active Design Modules: ADT, MLH, TRIZ, QFD Primary Active AI Modules: e.g. DRS, FLS
CDS/KJE (S1/Block 3)	*Justification* of the resulting selection of DPs recommended to designer to fulfill the FRs within constraints	MVC paradigm *Model* – Provides a systematic basis for the operations *View* – Provides transparency to the operations of the model *Controller* – Provides a means to interact with the model *Visualization Display*

EMBODIMENT DESIGN STAGE RELIABILITY AUTOMATION

The Embodiment Design Stage (EDS) supports the *development* of structural layout and component issues, predominantly in terms of *quantitative* analysis and assessment. We will discuss the reliability automation of the engine block modules during EDS, where we show the functions (Table 3.20) and active engines (Table 3.21).

ACTIVE RELIABILITY SUPPORT MODULES IN EDS BLOCK ENGINES

EDS/KAE Block 1

The embodiment design stage (EDS) KAE engine is responsible for assimilating *quantitative* knowledge by collecting, gathering, and combining knowledge from user and environment. This block monitors designer activity by providing templates for engineering input on various design layout and component options in preparation for processing in EDS/KCE. This may include monitoring the world for new layouts and components (marked up in OWL ontologies) that could be used in the design project based on user input in preparation for processing in KCE. Domain expert input can be accomplished using ontology editors such as CmapTools or Protege and then made machine-processible by exporting directly into OWL.

EDS/KCE Block 2

The EDS/KCE is responsible for coordinating, tasking, and processing the influx of knowledge from the KAE with the AI design engine block. In this block, for instance, systems engineering tasks at the component and structural level can be conducted, such as design structure matrix for component-component interaction, reliability block diagram (RBD) analysis for calculating series and parallel reliability, and fault tree analysis (FTA) to pinpoint faults. Domain rule support, predicate logic support, and fuzzy logic support can be provided as supportive inference mechanisms.

EDS/KJE Block 3

The EDS/KJE is responsible for justifying the automated *quantitative* recommendations by providing an interaction mechanism using an MVC, justifiable recommendations, and a rich visualization display. In this block, the output can be provided by a standard MVC connected to a Java Applet that enables interactive GUI display. The display can show problem areas in the design showing high risk with a color code corresponding to the numerical weight calculated by the design modules.

Table 3.20. EDS implementation schema.

Embodiment Stage	Purpose	Implementation Schema
EDS/KAE (S2/Block 1)	To assimilate *quantitative* knowledge by collecting, gathering, and combining knowledge from user and environment.	Monitoring OBIT designer activity by providing templates for engineering input on various design layout and component options in preparation for processing in KCE
EDS/KCE (S2/Block 2)	To correlate *quantitative* knowledge by coordinating, tasking, and processing the influx of knowledge from the KAE with the AI design engine block	Component to component analysis for reliability engineering can be done with techniques such as DSM, RBD, FTA
EDS/KJE (S2/Block 3)	To justify the automated *quantitative* recommendations by providing an interaction mechanism using an MVC, justifiable recommendations, and a rich visualization display	Show output using a standard MVC connected to a Java Applet that enables interactive GUI display. The display can show problem areas in the design showing high risk with a color code corresponding to the numerical weight calculated by the design modules quantitatively

Table 3.21. Embodiment design stage engines.

Stage/Engine	Processing Function	Active Region
EDS/KAE (S2/Block 1)	Assimilation of structural elements of design such as engineering layout and components	Monitoring of User Activity Monitoring of World Activity
EDS/KCE (S2/Block 2)	Correlation of structural elements and constraints with appropriate support engines in a way so that the designer's activity is supported by CommonKADS framework for KBE	Primary Active Task: Analytic Tasks, Synthetic Tasks Primary Active Design Modules: DSM, TRF, FTA, RBD, FMEA Primary Active AI Modules: DRS, PLS
EDS/KJE (S2/Block 3)	Justification of the resulting selection of layout and components	MVC paradigm *Model* – Provides a systematic basis for the operations *View* – Provides transparency to the operations of the model *Controller* – Provides a means to interact with the model *Visualization Display*

DETAIL DESIGN STAGE RELIABILITY AUTOMATION

The Detail Design Stage (DDS) supports the *development* of structural layout and component issues, predominantly in terms of *quantitative* analysis and assessment. We will discuss the reliability automation of the engine block modules during DDS, showing the function (Table 3.22) and active engines (Table 3.23)

ACTIVE RELIABILITY SUPPORT MODULES IN DDS BLOCK ENGINES

DDS/KAE Block 1

The EDS/KAE is responsible for assimilating *both qualitative and quantitative* knowledge by collecting, gathering, and combining knowledge from user and environment. In this block, monitoring designer activity is done by providing templates for engineering input on preferred options for design optimization in preparation for processing in DDS/KCE. Also, monitoring the world is done for new layouts and components (marked up in OWL ontologies) that could be used in the design project based on user input in preparation for processing in DDS/KCE. Input is by DE done using CmapTools and then made machine-processible using Visual OWL software.

DDS/KCE Block 2

The DDS/KCE is responsible for assimilating *both qualitative and quantitative* knowledge by collecting, gathering, and combining knowledge from user and environment. The Primary Active Task implementation could be used for risk mitigation tasks, e.g. *Analytic/Assessment Synthesis/Configuration.* Primary active design modules are FMEA, ETP, OPT such that when the FMEA module is executed, calculations are made considering the quantitative weights and the qualitative failure modes for the components, at which time the total entropy of the system can be assessed to determine the organizational status of the design, i.e. is it going towards disorder or order (for risk mitigation), furthermore, the OPT module can be use to assess MTTF and all types of cost and reliability optimizations in real-time as design progresses, e.g. efficiency can be maximized to determine best number of product engineering stages. The Primary Active AI Modules, e.g. CTS, DRS, and PLS, since the CTS can be utilized for keeping records of individual component simulations at T=1 to n=100 hours, for instance, and then assess the effect of clustering in different configuration on the component performance.

DDS/KCE Block 2

The DDS/KCE is responsible for justification of the resulting selection of blue

print configuration for design risk mitigation. In this block, the output is shown using a

standard MVC connected to a Java Applet that enables interactive GUI display, with

color-coding for risk areas as before.

Table 3.22. Detailed design implementation schema.

EDS Engines	Purpose	Implementation Schema
DDS/KAE (S3/Block 1)	To assimilate *both qualitative and quantitative* knowledge by collecting, gathering, and combining knowledge from user and environment	Monitoring designer activity by providing templates for engineering input on preferred options for design optimization in preparation for processing in KCE…Monitoring the world for new layouts and components (marked up in OWL ontologies) that could be used in the design project based on user input in preparation for processing in KCE
DDS/KCE (S3/Block 2)	To correlate *both qualitative and quantitative* knowledge by coordinating, tasking, and processing the influx of knowledge from the KAE with the AI design engine block	Primary active design modules are FMEA, ETP, OPT such that when the FMEA module is executed, calculations are made considering the quantitative weights and the qualitative failure modes for the components, at which time the total entropy of the system can be assessed to determine the organizational status of the design,
DDS/KJE (S3/Block 3)	To justify the *both* automated *qualitative and quantitative* recommendations by providing an interaction mechanism using an MVC, justifiable recommendations, and a rich visualization display	In this block, the output is shown using a standard MVC connected to a Java Applet that enables interactive GUI display, with color-coding for risk areas as before.

Table 3.23. Detailed design stage active engines.

Stage/Engine	Processing Function	Active Region
DDS/KAE (S3/Block 1)	Assimilation of optimized elements of design such as configuration possibilities	Monitoring of User Activity Monitoring of World Activity
DDS/KCE (S3/Block 2)	Correlation of optimized elements and constraints with appropriate support engines in a way so that the designer's activity is supported by CommonKADS framework for KBE	Primary Active Task: Analytic Tasks, Synthetic Tasks Primary Active Design Modules: FMEA, ETP, OPT Primary Active AI Modules: CTS, DRS, PLS
DDS/KJE (S3/Block 3)	Justification of the resulting selection of blue print configuration for design risk mitigation	MVC paradigm *Model* – Provides a systematic basis for the operations *View* – Provides transparency to the operations of the model *Controller* – Provides a means to interact with the model *Visualization Display*

ENABLING FRAMEWORK CONFIGURATION

IDENTIFYING REAL-WORLD NEED TO DEVELOP ENGINE BLOCK MODULES

Identifying RWN in Design Engine Block

Before identifying the modules in the design engine block, the real-world needs (RWN) had to be assessed first (Table 3.24) and mapped to modules (Table 3.25) after identifying them based on established design techniques, generally considered valid approaches to design independent of automation. Each of these RWN are essential for traceability of the functional requirements to the design parameters, and finally to the process variables, which actually define how the design parameters are acquired or fabricated. Thus, a sound foundation developed using axiomatic principles during architecture-driven software engineering is important before automation.

Table 3.24. Defining RWN mapped to design engine modules.

RWN	Module Function
RWN-1	Automated mechanism needed for hierarchical decomposition of FR and DP
RWN-2	Automated mechanism needed for invention
RWN-3	Automated mechanism needed for going from axiomatic analysis to components
RWN-4	Automated mechanism needed to ensure the customer guidelines are included in the quality of the design
RWN-5	Automated mechanism needed to determine component to component interaction
RWN-6	Automated mechanism needed to predict component failures
RWN-7	Automated mechanism needed to estimate system reliability
RWN-8	Automated mechanism needed to assess fault qualitatively and trace root cause
RWN-9	Automated mechanism needed to assess individual component risk on a cluster, mainly by associating any given component with a multiplier that affects the DSM
RWN-10	Automated mechanism needed to assess level of disorganization in system design during design process
RWN-11	Automated mechanism needed to manipulate domain-specific knowledge for inference, specifically in field of optical backplane engineering

Table 3.25. Defining RWN of AI mapped to inference engine modules.

RWN	Module Function
RWN-1	An automated mechanism need for domain rule support
RWN-2	An automated mechanism need for logic support
RWN-3	An automated mechanism need for algorithmic reasoning support
RWN-4	An automated mechanism need for heuristic reasoning support
RWN-5	An automated mechanism need for machine learning support
RWN-6	An automated mechanism need for advanced search algorithm support
RWN-7	An automated mechanism need for component influence assessment support
RWN-8	An automated mechanism need for weighted belief network analysis
RWN-9	An automated mechanism need for advanced data retrieval support

ASSIGNING DESIGN PARAMETERS TO FUNCTIONAL REQUIREMENTS

Design Modules as Design Parameters for Design Engine Block

For the purpose of configuration analysis, the AIDF modules are assigned axio-matic nomenclature (DP1, DP2, DP3,…DP11) for eleven design modules (Table 3.26).

Table 3.26. Assignment of modular DPs to fulfill FR real-world needs.

Design Parameters	Design Modules	Acronym	Module Function (refined functional requirement)
DP1	ADT	Axiomatic Design Theory	Provides an automated mechanism for hierarchical decomposition of FR and DP, provides 2 axioms, 11 corollaries, and 23 theorems for the rules base stored in the AI Engine Block
DP2	TRIZ	Theory of Inventive Problem Solving	Provides an automated mechanism for invention, especially by searching the Semantic Web for appropriate DPs
DP3	MLH	Hierarchical Multi-layer [design]	Providing an automated mechanism for going from FR to DP to components, calculation of reliability and cost
DP4	QFD	Quality Function Deployment	Provides an automated mechanism to ensure the customer guidelines are included in the quality of the design
DP5	DSM	Design Structure Matrix	Provides an automated mechanism to determine component to component interaction
DP6	FTA	Fault Tree Analysis	Provides an automated mechanism to predict component failures, where the calculations are based heavily on quantitative Boolean operators
DP7	RBD	Reliability Block Diagram	Provides an automated mechanism to estimate system reliability, where the calculations are based heavily on various engineering equations, such as MTTF (Mean time to Failure)
DP8	FMEA	Failure Mode Effects Analysis	Provides an automated mechanism to associate weights for each type of failure to assess fault qualitatively and trace root cause
DP9	TRF	Technology Risk Factor	Provides an automated mechanism to assess individual component risk on a cluster, mainly by associating any given component with a multiplier that affects the DSM
DP10	ETP	Entropy	Provides an automated mechanism to assess level of disorganization in system design during design process
DP11	OPT	Optical Backplane Engineering Domain	Provides an automated mechanism to manipulate domain-specific knowledge for inference, specifically in field of optical backplane engineering

Inference Modules as Design Parameters for AI Design Engine Block

For the purpose of configuration analysis, the AIDF modules are assigned axio-matic nomenclature (DP1, DP2, DP3,…DP9) for nine inference modules (Table 3.27).

Table 3.27 Assigning DP to AIDF inference engine module functions.

Design Parameters	AI Modules	Acronym	Module Function
DP1	DRS	Domain Rule Support	Provides an automated mechanism for domain rule support, in terms of executable rules used by the inference engine
DP2	PLS	Predicate Logic Support	Provides an automated mechanism for logic support
DP3	ARS	Algorithmic Reasoning Support	Provides an automated mechanism for miscellaneous algorithmic reasoning support
DP4	FLS	Fuzzy Logic Support	Provides an automated mechanism for fuzzy logic support
DP5	NNS	Neural Network Support	Provides an automated mechanism for neural network support
DP6	GAS	Genetic Algorithm Support	Provides an automated mechanism for genetic algorithm support
DP7	CTS	Conant Transmission Support	Provides an automated mechanism for component transmission support
DP8	CBS	Calibrated Bayesian Support	Provides an automated mechanism for calibrated Bayesian support
DP9	DMS	Data Mining Support	Provides an automated mechanism for data mining support

ENABLING MATRIX DECOMPOSITION FOR ALL STAGES

Functional Requirement Design Engine Block Decomposition for All Stages

One can hierarchically decompose the design engine block starting with the top-level module identified, where each FR can be further broken down into hundreds of constituent elements. This process of axiomatic design ensures that each need is accounted for by at least one module, and allows for connecting between elements between subelements of each module for detailed configuration. Each design module DP fulfills the need for each FR, so we can construct a FR/DP design matrix for design modules (Table 3.28).

Table 3.28. Design engine modules FR/DP design matrix.

	Modules	DP1	DP2	DP3	DP4	DP5	DP6	DP7	DP8	DP9	DP10	DP11
FR1	ADT	X										
FR2	TRIZ		X									
FR3	MLH			X								
FR4	QFD				X							
FR5	DSM					X						
FR6	FTA						X					
FR7	RBD							X				
FR8	FMEA								X			
FR9	TRF									X		
FR10	ETP										X	
FR11	OPT											X

Inference Engine Block Decomposition for All Stages

One can hierarchically decompose the inference engine block starting with the top-level module identified, where each FR can be further broken down into hundreds of constituent elements. This process of axiomatic design ensures that each need is accounted for in each module, and allows for connecting between elements between subelements of each module for detailed configuration. We will describe the configuration process in more detail in the case study chapter. Each inference module DP fulfills the need for each FR, so we can construct a FR/DP design matrix for inference modules (Table 3.29).

Table 3.29. Inference engine modules FR/DP design matrix.

		DP1	DP2	DP3	DP4	DP5	DP6	DP7	DP8	DP9
FR1	DRS	X								
FR2	PLS		X							
FR3	ARS			X						
FR4	FLS				X					
FR5	NNS					X				
FR6	GAS						X			
FR7	CTS							X		
FR8	CBS								X	
FR9	DMS									X

ENABLING MODULE CONFIGURATION PER STAGE

In this section, we are interested in the active modules in each stage. We will show that each of the twenty functional requirements is fulfilled by at least one design parameter corresponding to a module in the engine block. After showing how to set up a framework enabling configuration, we will show the configuration process in the case study chapter in more detail.

CDS/KCE (S1/Block 2) Engine-Block Active Modules

The active tasks during conceptual design support are primarily analytic tasks. In this stage, various analytic subtasks will activate the design engine modules identified as ADT, MLH, TRIZ, and QFD and inference engine block modules identified as DRS, FLS (Table 3.30). In the conceptual design stage, the engine is involved in correlation of FRs, DPs, and constraints with appropriate support engines in a way so that the designer's activity is supported by CommonKADS framework for KBE. The corresponding representation as a design matrix matching FR with corresponding DP is shown separately for the design engine (Table 3.31) and inference engine (Table 3.32).

Table 3.30. Stage I active modules.

CDS/KCE (S1/Block 2)	*Correlation* of FRs, DPs, and constraints with appropriate support engines in a way so that the designer's activity is supported by CommonKADS framework for KBE	Primary Active Task: Analytic Tasks Primary Active Design Modules: ADT, MLH, TRIZ, QFD Primary Active AI Modules: e.g. DRS, FLS

Table 3.31. Stage I active design engine modules FR/DP design matrix.

		DP1	DP2	DP3	DP4	DP5	DP6	DP7	DP8	DP9	DP10	DP11
FR1	*ADT*	X										
FR2	*TRIZ*		X									
FR3	*MLH*			X								
FR4	*QFD*				X							
FR5	DSM											
FR6	FTA											
FR7	RBD											
FR8	FMEA											
FR9	TRF											
FR10	ETP											
FR11	OPT											

Table 3.32. Stage I inference engine modules FR/DP design matrix.

		DP1	DP2	DP3	DP4	DP5	DP6	DP7	DP8	DP9
FR1	*DRS*	X								
FR2	PLS									
FR3	ARS									
FR4	*FLS*				X					
FR5	NNS									
FR6	GAS									
FR7	CTS									
FR8	CBS									
FR9	DMS									

EDS/KCE (S2/Block 2) Engine-Block Active Modules

The active tasks during embodiment design support are both analytic and synthetic tasks. In this stage, various analytic and synthetic subtasks will activate the design engine modules identified as DSM, TRF, FTA, RBD, and FMEA and inference engine modules identified as DRS, PLS (Table 3.33). In the embodiment design stage, the engine is involved in correlation of structural elements and constraints with appropriate support engines in a way so that the designer's activity is supported by CommonKADS framework for KBE. The corresponding representation as a design matrix matching FR with corresponding DP is shown separately for the design engine (Table 3.34) and infer-

ence engine (Table 3.35).

Table 3.33. Stage II active modules.

EDS/KCE (S2/Block 2)	Correlation of structural elements and constraints with appropriate support engines in a way so that the designer's activity is supported by CommonKADS framework for KBE	Primary Active Task: Analytic Tasks, Synthetic Tasks Primary Active Design Modules: DSM, TRF, FTA, RBD, FMEA Primary Active AI Modules: DRS, PLS

Table 3.34. Stage II active design engine modules FR/DP design matrix.

		DP1	DP2	DP3	DP4	DP5	DP6	DP7	DP8	DP9	DP10	DP11
FR1	ADT											
FR2	TRIZ											
FR3	MLH											
FR4	QFD											
FR5	*DSM*					X						
FR6	*FTA*						X					
FR7	*RBD*							X				
FR8	FMEA											
FR9	*TRF*									X		
FR10	ETP											
FR11	OPT											

Table 3.35. Stage II inference engine modules FR/DP design matrix.

		DP1	DP2	DP3	DP4	DP5	DP6	DP7	DP8	DP9
FR1	*DRS*	X								
FR2	*PLS*		X							
FR3	ARS									
FR4	FLS									
FR5	NNS									
FR6	GAS									
FR7	CTS									
FR8	CBS									
FR9	DMS									

169

DDS/KCE (S3/Block 2) Engine-Block Active Modules

The active tasks during embodiment design support are primarily synthetic tasks. In this stage, various analytic and synthetic subtasks will activate the design engine modules identified as FMEA, ETP, OPT and inference engine modules identified as CTS, DRS, PLS (Table 36). In the embodiment design stage, the engine is involved in correlation of structural elements and constraints with appropriate support engines in a way so that the designer's activity is supported by CommonKADS framework for KBE. The corresponding representation as a design matrix matching FR with corresponding DP is shown separately for the design engine (Table 3.37) and inference engine (Table 3.38).

Table 3.36. Stage III active modules.

DDS/KCE (S3/Block 2)	Correlation of optimized elements and constraints with appropriate support engines in a way so that the designer's activity is supported by CommonKADS framework for KBE	Primary Active Task: Analytic Tasks, Synthetic Tasks Primary Active Design Modules: FMEA, ETP, OPT Primary Active AI Modules: CTS, DRS, PLS

Table 3.37. Stage II active design engine modules FR/DP design matrix.

		DP1	DP2	DP3	DP4	DP5	DP6	DP7	DP8	DP9	DP10	DP11
FR1	ADT											
FR2	TRIZ											
FR3	MLH											
FR4	QFD											
FR5	DSM											
FR6	FTA											
FR7	RBD											
FR8	*FMEA*								X			
FR9	TRF											
FR10	*ETP*										X	
FR11	*OPT*											X

Table 3.38. Stage II inference engine modules FR/DP design matrix.

		DP1	DP2	DP3	DP4	DP5	DP6	DP7	DP8	DP9
FR1	*DRS*	X								
FR2	*PLS*		X							
FR3	ARS									
FR4	FLS									
FR5	NNS									
FR6	GAS									
FR7	*CTS*							X		
FR8	CBS									
FR9	DMS									

ENGINE BLOCK CONFIGURATION DEPENDENCY MATRIX

We will show how to configure this matrix to adapt to the needs of various domains, e.g. some application domains will require the operation of multiple modules to fulfill one functional requirement. Already it has been shown that various stages in the design process have different active modules, which can be depicted with this matrix as well. In this section, we will prepare the AIDF engine block (Fig. 3.32) for configuration with a generic matrix showing a one-to-one mapping of FR to DP. We will discuss modular interaction analysis in the engine block, as well as the interlacing dual engine operations for robust support. In the generic framework, we enable configuration to occur for each engine block independently for design engine and inference engine. We also introduce the concept of the AIDF providing a "hybrid" system when both engine blocks are operating together, achieved when the modules in the design engine block are interlaced together with the inference engine block modules. This enables more engine blocks to operate cohesively in parallel, which can be expanded in a process similar to modular expansion of each engine. The AIDF currently has two engines, but, in theory, there is no limit to the number of engines and modules we can introduce.

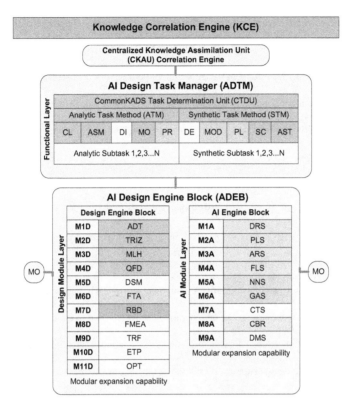

Fig. 3.32. Knowledge correlation engine framework for configuration.

MODULAR INTERACTION ANALYSIS IN ENGINE BLOCK

The engine block operation can be interlaced together. Based on the design modules in the design engine block, one can construct a design matrix representation of the dependencies, or interactions, for a given domain. Furthermore, if there is coupling between engines, between the design engine and inference engine, that configuration also can be represented to develop different applications for various domains emphasizing different calculations. In the case study chapter, we will scope out the Reliability (RBD) module for interaction analysis, including a framework to show speed and design matrix optimization as part if pre-deployment DFSS IDOV validation, followed by showing how the AIDF fits into the AIDF GAURDS validation framework post-deployment. A comprehensive validation methodology for the architecture and other aspects of a KBE SoS are provided in the validation chapter. In the generic framework, we enable configuration to occur for each engine block independently for design engine and inference engine.

INTERLACING DUAL ENGINE OPERATIONS FOR ROBUST SUPPORT

The AIDF has a dual engine block in the KCE that can operate independently as a design (Table 3.39) or inference engine (Table 3.40) or in tandem (Table 3.41). The data flow of the modules and their interlacing interaction with each other and with Web Services can be observed (Fig 3.33), which we will show in detail in the case study chapter after actual configuration. This capability allows it function strictly as a traditional KBE system providing design support, if only the design engine block modules are enabled for configuration, or function as an "expert" system relying only on inference support, in only the inference engine block modules are enabled for configuration. Thus, the AIDF provides a "hybrid" system when both engine blocks are operating together achieved when the modules in the design engine block are interlaced together with the inference engine block modules. We introduce a framework to connect the two engine blocks together for interlacing support of their operation. A configured example is provided in the case study chapter.

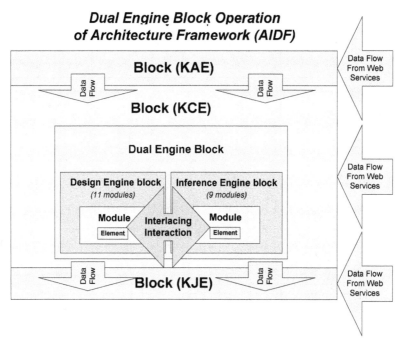

Fig. 3.33. Intended data flow of the generic AIDF platform before configuration.

Table 3.39. Framework for identifying interacting design modules.

	Modules	DP1	DP2	DP3	DP4	DP5	DP6	DP7	DP8	DP9	DP10	DP11
DP1	ADT	X										
DP2	TRIZ		X									
DP3	MLH			X								
DP4	QFD				X							
DP5	DSM					X						
DP6	FTA						X					
DP7	RBD							X				
DP8	FMEA								X			
DP9	TRF									X		
DP10	ETP										X	
DP11	OPT											X

Table 3.40. Framework interlacing interacting inference modules for configuration.

		DP1	DP2	DP3	DP4	DP5	DP6	DP7	DP8	DP9
DP1	DRS	X								
DP2	PLS		X							
DP3	ARS			X						
DP4	FLS				X					
DP5	NNS					X				
DP6	GAS						X			
DP7	CTS							X		
DP8	CBS								X	
DP9	DMS									X

Table 3.41 Framework interlacing dual engine block operations for configuration.

#	DP	Design/Inference Modules	1	2	3	4	5	6	7	8	9	10	11	12	13	14	15	16	17	18	19	20
1	DP1	ADT	X																			
2	DP2	TRIZ		X																		
3	DP3	MLH			X																	
4	DP4	QFD				X																
5	DP5	DSM					X															
6	DP6	FTA						X														
7	DP7	RBD							X													
8	DP8	FMEA								X												
9	DP9	TRF									X											
10	DP10	ETP										X										
11	DP11	OPT											X									
12	DP1	DRS												X								
13	DP2	PLS													X							
14	DP3	ARS														X						
15	DP4	FLS															X					
16	DP5	NNS																X				
17	DP6	GAS																	X			
18	DP7	CTS																		X		
19	DP8	CBS																			X	
20	DP9	DMS																				X

DISCUSSION OF ARCHITECTURE DEVELOPMENT STRATEGY

DISCUSSION OF FRAMEWORK FEATURES

In addition to its validated *integrity*, the AIDF has been developed to have a high degree of *usability* and *maintainability* due to its model-driven architecture (MDA) approach that allows technology independence over time. The framework's domain-specific *configurability* can be attributed to its modular engine block housing 20 modules configurable for AI and Design. The high-configurability of the AIDF is attributed to the *modular* framework structure, allowing as many modules to be added and activated for design and AI support in the engine block. In this thesis, we address twenty unique modules capable of being configured to support a domain-specific design environment such as optical backplane engineering. Each module can be *knowledge-engineered* according to stated needs for inductive and deductive reasoning, as well as brute-force design calculations. This process of knowledge engineering is a method by which structural and dynamic models can be developed using SysML and UML to define algorithmic processes occurring in a given module. All types of process configurations can be designed for each module that conduct its respective analysis and calculations, such as making design calculations or invoking procedural calls from templates continuously being updated by domain experts situated world-wide. For instance, intelligent agents can be directed to retrieve information from authenticated knowledge repositories on the Semantic Web to provide proactive recommendations on state-of-the-art components in industrial use to the optical backplane designer. We successfully tested how domain experts will be able to conveniently mark-up their domain-specific components using an ergonomic visual ontology language editor such as Concept Map Tools (CmapTools), with little or no learning curve. Then we successfully tested *exportability* of the knowledge base by checking to see if we could really *export* our created ontologies into eXtensible Markup Language and Web Ontology Language. As a result, we were able to demonstrate a *scalable* case study example of how to develop a knowledge base using ontologies for free-space optical interconnect technology having on the order of a dozen components. More importantly, we determined that it will be possible to scale up for an actual KBE system implementation having thousands of components that an inference engine will work on, both locally and remotely. The *local* knowledge base can be protected with a firewall for *security*, whereas the *remote* knowledge repositories accessed by intelligent agents dispatched by the AIDF KBE system will need to be authenticated with mechanisms such as Coda. Furthermore, we provide a methodology for validation of the ontologies using current research showing how to evaluate mathematically the quality of the ontologies in terms of their factors for breadth, fan-out, and tangledness.

DISCUSSION OF ARCHITECTURAL DEVELOPMENT PRINCIPLES

After a survey of various architectures already in use, such as U.S. DoD C4ISR, Zachman Framework, ISO RM-ODP [Maier and Rechtin, 2000], we converged on adopting the terminology used for definitions in *ANSI/IEEE Std. 1471-2001* and Carnegie Mellon's Software Engineering Institute definitions being formulated by Shaw and others after during and after the *First International Workshop on IT Architectures for Software Systems in 1995* [Garlan, 1995a]. In April 1995, the IEEE Software Engineering Standards Committee (SESC) convened an Architecture Planning Group (APG) to study the development of an architecture standard for software-intensive systems [IEEE-1471, 2006]. After publication of their report, the APG upgraded to the Architecture Working

Group and was charged with the development of a Recommended Practice for architectural description – a particular type of standard. Hence, the framework model of the AIDF uses the current *ANSI/IEEE Std. 1471-2000* terminology that is consistent with the definition of architecture, as defined by Congress in the *National Defense Authorisation Act for Fiscal Year 1996* [Clinger-Cohen Act, 1999]. The architectural model categorizations of architecture, namely framework model, structural model, dynamic model, process model, and sometimes functional model have been adopted this thesis, based on the results of Shaw's compilation of the submissions to the *First International Workshop on IT Architectures for Software Systems in 1995* [Shaw, M., D. Garlan, 1995]. The technological independence of the AIDF model is achieved by applying a very recent architectural trend that ensures long-term maintainability of a system by application of Object Management Group's *Model Driven Architecture (MDA)* approach to system design. Furthermore, the three phase design process for which the AIDF provides corresponding three-stage support is based on the established engineering standards of what constitutes the major phases of the design process, namely conceptual, embodiment, and design phases [Pahl and Beitz, 1988]. The AIDF architectural framework that provides support for each phase has a partitioning strategy based on modularity allowing compartmentalization of functional concerns. This enables implementation of the AIDF architectural framework suitable for use with Acclaro DFSS that captures the functional requirements of each stage and maps it to appropriate design parameters that fulfill each requirement. The expansion capability of the AIDF architectural framework's AI design engine is attributed to the *modular* approach. A review of the Architecture Tradeoff Analysis Model (ATAM) [Bass, et al., 2003] and the Cost Benefit Analysis Method (CBAM) [Bass, et al.,

2003] provided insight on some of the most current and systematic approaches to architecture analysis and development.

The benefits of this approach implemented with Acclaro DFSS, using the axiomatic approach to the design process, is that validation techniques can be introduced at the outset of requirements specifications (front-end) instead of having to wait until deployment, which could be too late to make changes or simply too costly. Hence, the fundamental two design axioms, namely the independence axiom for assessing design coupling and the information axiom for assessing design complexity can immediately be applied to get an idea on the appropriateness of the KBE solution first-hand in both a qualitative and quantitative way. Furthermore, the associated 11 design corollaries and 23 theorems can be applied systematically to the functional requirements and design parameters captured, while creating the design matrix. Although this tool alone was sufficient to meet the validation needs of the architectural framework in Tier-1, for completeness sake, we continued to research a cascade of tools that could be used in conjunction with Acclaro DFSS all the way down to Tier 5 validation methodologies.

DISCUSSION OF HIGH LEVEL INTEGRATED DESIGN FOR GAURDS

In order to build the "right" KBE system, we periodically consulted with the Artificial Intelligence group at NASA Marshall Space Flight Center during the two-year Fellowship Training Grant to get advice on what constituted a valid architectural framework and surveyed the field on the state-of-the-art for validation in general, only to discover that validation itself is an art form that is emerging as a discipline [Hoppe, 1993]. In our case, we developed an architecture framework that can be validated before deployment, as well as after deployment. Our focus is on validation before deployment, although we have provided a methodology to follow after deployment as well, in order to keep pace with advances in technology. Therefore, we used Acclaro DFSS to capture the high-level requirements specifications in the form of functional requirements and design parameters to produce the AIDF architectural framework that meets the needs of our particular domain-specific OBIT KBE system's expressed and evolving needs, which was combined with GAURDS validation framework to ensure we have a methodology to validate the entire system, i.e. to ensure the design of the "right" system.

The actual development and deployment of a KBE system based on the AIDF would require the efforts of a team of engineers over the course of years for a particular domain, such as OBIT, with twenty separate modules provided by the AIDF knowledge engineering methodology. In order to grapple with this validation task, we searched for a direct, industry-grade, mission-critical approach to validate a high level architectural framework before deployment that was more appropriate for the scope of a thesis dissertation and found an industrial grade system already in use for construction of nuclear submarines, space systems, and railways called the GAURDS validation framework. Furthermore, we discovered that peripheral support provided by High Level Integrated Design for dependability could be used to validate the structural and dynamic models of the object-oriented design of the AI Design Engine. Hence, we decided to make the centerpiece of our approach GAURDS, which stands for Generic Architecture for Upgradeable Real-time Systems with support from HIDE. However, before we applied GAURDS validation framework, we had to develop a valid approach to development of the architectural framework for the AIDF itself.

DISCUSSION OF ARCHITECTURE TRADEOFF ANALYSIS MODEL

The Software Engineering Institute's (SEI) Architecture Tradeoff Analysis

Method (ATAM) (Fig. 3.34) is one method for evaluation of software architecture. An

evaluation based on the ATAM typically takes three to four days of group analysis com-

prised of an evaluation team, architects, and representatives of the architecture's various

stakeholders. Some benefits of the ATAM include clarified quality attribute require-

ments, improved architecture documentation, documented basis for making architectural

decisions, the ability to identify risks, and increased communication among stakeholders.

The following diagram displays a conceptual flow of the ATAM. The business drivers

and the software architecture are elicited from discussions with project decision-makers.

In the case of the AIDF architecture development, the "business drivers" were more ori-

ented to the industrial needs of the case study, and the software architecture was based on

Shaw's school of thought. These are refined into scenarios and the architectural decisions

made in support of each one.

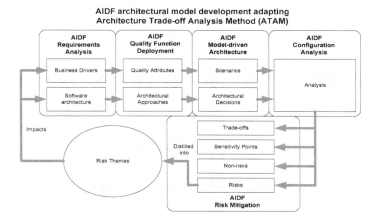

Fig. 3.34. Conceptual flow of ATAM for architectural trade-off analysis.

Many of the risks and non-risks, tradeoff possibilities, and sensitivity points of the architecture were revealed by scenario analysis techniques. Some example scenarios that strengthened the AIDF architectural framework were involved considering issues such as how to build in features, such as performance attributes, that would address all types of possible situations that may arise during KBE system operation: these include scenarios such as a rule in the rule engine fires with slow reaction time in the rule-base; a domain expert provides knowledge asynchronously for human convenience that the intelligent agents must detect and retrieve automatically at any time; a multi-lingual knowledge base support; authentication of the knowledge base in the continuously updated Web repository; local and remote knowledge base querying time and quality; on demand reporting by the designer needing justification of the design recommendations throughout each stage of the design process support. The most important results of the brainstorming of these plausible scenarios during development also helped in the process of improvement for the AIDF architectural framework, as well as in development of the AIDF-SVM synergistic validation methodology. The various strategies and tactics for analysis of scenarios and decisions help to identify risks, non-risks, sensitivity points, and tradeoff points in the architecture. Many of the identified architectural risks are synthesized into a set of risk themes, showing how each one could threaten a business driver. The type of scenarios we considered were based on those encountered in optical backplane engineering, while considering the impact on all engineering disciplines for broad impact.

Trade-offs for various attributes during the development process of the AIDF architectural framework model were addressed in part by using ATAM architectural evaluation methods. For instance, a balance between speed of data access and quality of knowledge was resolved by making sure that a local knowledge base existed for rapid access, with a remote and distributed knowledge base could co-exist on a network accessible asynchronously using intelligent agents acting on and traversing the Semantic Web, searching for the most recent pieces of component and industrial knowledge. We considered many features in the development of the AIDF architectural framework that would improve on system availability, modifiability, performance, security, testability, and usability, with some assistance from ATAM for evaluation and trade-off issues. In order to achieve AIDF KBE system *availability*, we considered fault detection, recovery, repair, and prevention. In order to achieve AIDF KBE system *modifiability*, we considered features such as localized changes, prevention of ripple effects, and deferred binding time by introducing a modular engine block, OMG MDA approach and five-tier validation approach. In order to achieve AIDF KBE system *performance*, we considered features such as resource demand that would increase computational efficiency and reduction of computational overhead, resource management by introducing concurrency of operations for each of the three blocks comprising the AIDF, and increase of available resources knowledge repositories available by leveraging Semantic Web Services. In order to achieve AIDF KBE system *testing* ability, we developed a comprehensive 5-tier validation approach within the AIDF-Synergistic Validation Methodology (SVM), explained in detail in the validation chapter. In order to achieve AIDF KBE system *usability*, we developed the concept of having a separate user interface for domain experts, the systems administrator, and designer who uses the OBIT KBE system based on a model-view-controller architectural paradigm.

DISCUSSION OF COST BENEFIT ANALYSIS METHOD

Although ATAM serves as a guide for qualitative architectural decision-making and evaluation, the Cost Benefit Analysis Method (CBAM) (Fig. 3.35) builds on this foundation, providing a quantitative approach to architectural decision on assessing economic impact of trade-offs. The creation and maintenance of a complex software-intensive system based on the AIDF architecture involves making a series of critical architecture design decisions that would effect forthcoming KBE system development. ATAM provides software architects with a framework for understanding the technical tradeoffs encountered during development of architecture for KBE systems. However, some of the most significant tradeoffs in large complex systems also carry associated economic impact and repercussions that could last decades, after the software architecture guides deployment of actual KBE systems, which can be addressed by CBAM. Furthermore, ATAM does not necessarily provide much guidance for assessing such economic tradeoffs. Historically, when economics have been addressed in the past, the focus has usually focused on the costs of building the KBE system, not the long-term costs of maintenance, upgrade, and expansion. There are benefits to take time in making architectural decisions that could effect how hundreds of KBE systems will be built over time. Clearly we need to consider the Return on Investment (ROI) of any architectural decision since the resources for building and maintaining a system are not infinite. Thus, there must be a rational, quantifiable process for choosing among various architectural options. These options will have its associated costs and inherent risks, while consuming all types and amounts of resources. Hence, a method to explore the effects of these options, software models that emphasis economy are also needed to take into account the cost, benefits, risks, and schedule implications of critical architectural decisions. CBA M provides a

methodology to evaluate the cost impact of performance, availability, security, and modi-

fiability by applying quantitative methods for making calculated economic decisions in

terms of costs and benefits.

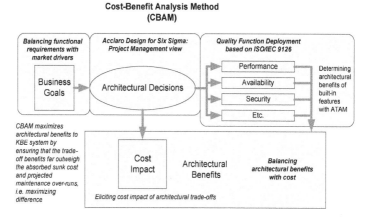

Fig. 3.35. CBAM for assessment of cost impact of architectural trade-off.

One of the applications of CBAM has been on the architectural cost ramifications

of the NASA Earth Observing System, which is a constellation of satellites that gathers

data about the Earth for the U. S. Global Change Research Program and other scientific

communities worldwide. The Software Engineering Institute (SEI) has applied CBAM on

Earth Observing System Data Information System (EOSDIS) Core System (ECS) with

the aim of making investment decisions for the project, for instance [Bass, et al., 2003].

By using the CBAM the ECS managers were able to order a set of architectural strategies

based upon their predicted Return On Investment (ROI). The CBAM process provided a

means to assess the potential impact of scenarios more clearly in advance, assigning utility levels to specific response goals, and to prioritize scenarios based on the resulting determination of utility and risk. Similarly, although the AIDF architectural framework is more directly involved in determining the functional requirements and their associated design parameters, ATAM and CBAM were useful brainstorming methods to evaluate various scenarios according to cost and risk issues. ATAM was used in brainstorming of various scenarios that helped define and redefine the FRs and DPs of the AIDF. CBAM was more useful in assessing the expected impact of the architectural decisions with respect to cost issues by prioritizing scenarios evaluated, such as potential maintenance costs, which led us to, for instance, building a modular engine block and outsourcing to knowledge repositories using Semantic Web Services to minimize on direct cost of acquiring memory allocation for the local knowledge base.

DISCUSSION OF VALID SOFTWARE ARCHITECTURE TERMINOLOGY

In order to build an AI system to support the design process, particularly a KBE design environment, that can that can stand the test of time, a suitable architecture must be developed, making the KBE system itself simply an instantiation of a particular framework for its structured and systematic development. Hence, the broad impact of the AIDF can be assessed by how many KBE systems can be developed based on its framework, which is limited only by the number of engineering disciplines and domains that the architectural framework can be applied to for guidance on KBE development. The actual number is virtually limitless, considering that the AIDF can be used to develop a KBE system that can be adapted to everything from optical backplane engineering to space systems design.

One of the first steps towards validation of architecture is to make sure that a commonly accepted foundation of definitions is established before continuing to more sophisticated forms of validation, verification and evaluation. Hence, after some research it was determined that IEEE 1471 terminology (Table 3. 42) is a good starting point, considering its popularity both in academia, as well as in industry. This move served as an important starting point for an appropriate approach to devising an all-encompassing meta-level functional architecture, which is expressed as a framework that is suitable for describing an AI design environment for any engineering domain. The conceptual build-ing blocks used later used for describing the AIDF were based primarily on this set of definitions for commonly used key words used in the literature, such as "model", "view", and "architecture".

Table 3.42. Architecture working group definition: IEEE 1471.

Title	Date	Source	Architecture Definition
ANSI/IEEE Std. 1471-2000	2000	IEEE	AWG Definition: *The fundamental organization of a system, embodied in its components, their relationships to each other and the environment, and the principles governing its design and evolution.* (Recommended Practice for Architectural De-scription of Software-Intensive Systems)

DISCUSSION OF VALID ARCHITECTURE SCHOOL OF THOUGHT

During the development phase of the architecture models for the AIDF, the effort spent researching and analyzing the many different schools of thought and approaches to architecture in the literature was critical as well as fruitful, considering the multitudes of competing definitions of what constituted suitable and proper architecture methodology. It was discovered that the U.S government had a legal definition of architecture that was consistent with Shaw's academic approach at Carnegie Mellon, which was accepted by the Software Community in the First International Workshop on IT Architectures for Software Systems (Table 3.43). Hence, this fact had bearing in selecting Shaw's categorization of architecture into five areas, laying the foundation for creation and further development of the AIDF.

Table 3.43. Comparing legal and academics architecture definitions.

Title	Date	Source	Definition
National Defense Authorisation Act for Fiscal Year 1996 (Clinger-Cohen Act)	1996	U.S. Congress	Legal Definition: Information Technology Management Reform Act of 1996 *An integrated framework for evolving or maintaining existing technology and acquiring new information technology to achieve the agency's strategic goals and information resource management.*
First International Workshop on IT Architectures for Software Systems	1995	International IT Workshop	Categorization: Five architectural models categorized and defined by Shaw: *Process Model* Provides framework development methodology of architecture *Framework Model* Provides an overall representation of architecture *Structural Model* Provides specific OO views expressing how each framework module is constructed using an ADL *Dynamic Model* Provides OO views expressing interactive and dynamic parts *Functional Model* Provides views expressing functionality

Consequently, a strategic decision was made to base the definitions of architecture according to Shaw's school of thought, established in the First International Workshop on IT Architectures for Software Systems. The benefits of this approach significantly helped in creating a robust architecture design for the AIDF by categorizing into five different types of models that show different aspects of a design.

DISCUSSION OF MODEL-DRIVEN ARCHITECTURE

The AIDF utilized Object Management Group's (OMG) model driven architecture (MDA) approach by building into the architectural framework mechanisms to adapt to long-term change in terms of years and even decades for a large-scale system requiring technological maintenance and platform flexibility. The three primary goals of MDA are portability, interoperability and reusability through architectural separation of concerns. Implementation, integration, maintenance, testing, and simulation of a KBE system are taken into account through the MDA approach [OMG, 2006].

Implementation of the infrastructure can be adopted by existing AIDF KBE systems by tracing the actual needs of the KBE system to its original set of high-level abstract functional requirements. *Integration* avenues are opened up by automation of the production of data integration bridges, as in the case of expansion of the KB of the AIDF by integration of modules. *Maintenance* methods are systematic, considering that the architectural blue-prints that are carefully prepared taking into account terminological conventions and clarifying high-level abstract associations of disparate parts of a system in machine-readable form make it possible for human developers to troubleshoot more systematically. *Testing and simulation* is made easier considering that the developed models can be used to generate code, as in the case of the structural model using Telelogic TAU for UML/SysML. In fact, these tools are even being equipped with a tool facility to validate and verify against the requirements originally provided in the framework model of Acclaro DFSS. These specifications are expected to lead the industry towards interoperable, reusable, portable software components, in addition to data models based on standard models.

The AIDF information technology infrastructure is effectively a distributed computing system that is heavily dependent on rapidly changing technology platform and implementation methods, based on the popular tools and programming techniques of the year. One way to make sure that the work done on an architectural framework that will persevere over time and survive the ravages of technological change on a system built today is to build it in a way that assures *technology independence*. Hence, a systems architect for any discipline, especially one with an intended impact on a myriad of domain specialties in the area of KBE, must recognize the need for adaptive growth of a system that does not affect the core functional requirements of the original architectural framework design. Hence, future generations can still rely on his efforts and not have to worry about legacy problems carried over from the past. The goal of using Model Driven Architecture in the AIDF architectural framework development process was predominantly to shift the phase of technology and platform selection as far out as possible, thereby assuring technology independence of the core architectural framework design.

DISCUSSION ON MODULARITY ASSURING UPGRADEABILITY

Published in October 2002, ISO/IEC 15288-System Life Cycle Processes is the first ISO standard to deal with system life-cycle processes approved and contributed to by INCOSE for hardware, software and human interfaces [ISO, 2002]. We use some of the ideas in this standard as guidance in upgrading the AIDF knowledge correlation engine (KCE). The framework of this standard encompasses the life cycle of man-made systems, which spans idea conception, maintenance, and the retirement of the system. ISO/IEC 15288 provides the processes for acquiring and supplying system products and services that are configured from one or more of the following types of system components:

hardware, software, and human interfaces, which we have adapted to the design and AI engine block modules of the AIDF.   The processes in this international standard form a comprehensive set from which an organization may construct life cycle models appropriate to the product, service types, and markets in which they trade, which is very suitable to the various domains that an AIDF KBE system may address.   This system life cycle process standard can provide valuable guidance in the design and maintenance of efficiently operating KBE systems. We revisit the KCE to show that standard can be applied in maintaining the engines of the AIDF (Fig. 3.47)

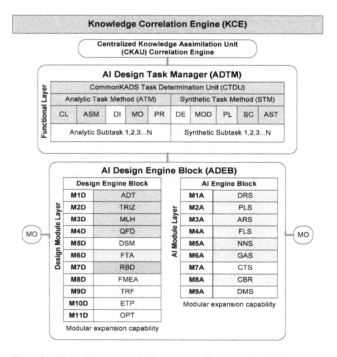

Fig. 3.36. Knowledge correlation engine framework detail.

SUMMARY

In this chapter, we provided the scope and impact of the AIDF. We provided a validation overview for the architecture framework. We provided a discussion of the chapter contents, before describing the AIDF input, processing, and output block that operate as the Knowledge Assimilation Engine (KAE), Knowledge Correlation Engine (KCE), and the Knowledge Justification Engine (KJE), respectively. We described each engine block in detail, providing snapshots of the implementation with Acclaro Design for Six Sigma. We described the operations in each block, followed by introducing the three stages of the AIDF corresponding to the three design phases. The operations in each design phase are introduced. A discussion of the architecture development strategy was provided. In the next chapter on case study, we will show the implementation, application, configuration, optimization, and verification of the case study based on optical backplane engineering.

CHAPTER IV: ARCHITECTURE FRAMEWORK CASE STUDY

OVERVIEW

In the case study chapter, we introduce the application, configuration, implementation, optimization, and verification, using the AIDF architecture framework for optical backplane engineering. In the application section, we describe the optical backplane components and how to apply knowledge engineering allowing domain experts to capture their tacit and explicit knowledge on components and design rationale, using methods such as web-enabled DSM templates, rules, and ontologies. In the configuration section, we show various levels of granularity in analyzing the AIDF, focusing at high-level down to engine module and element interaction analysis. In the implementation section, we provide screenshots using Acclaro DFSS design matrix architecture development tool. In the optimization and verification section, we show how the DFSS IDOV, combined with the GAURDS validation framework, provides a pre-deployment and post-deployment validation strategy for the AIDF architectural framework

AIDF ARCHITECTURE FRAMEWORK CASE STUDY APPLICATION

We have developed an architecture framework model for configuring a KBE SoS application for automated design and inference support for product engineering. Since our focus is on reliability engineering for design risk mitigation, we have selected the design of a Free-Space Optical Interconnect (FSOI) for AIDF configuration, due to its inherent design risks we wish to mitigate. Since the components of an FSOI operate at the micron level, it is critical that the vulnerabilities of these individual components and grouped component clusters are assessed and their inherent risk mitigated during the automated design process support. We will introduce the optical backplane components for a particular FSOI case study implemented in a laboratory in an actual optical backplane laboratory as a basis to configure the AIDF.

NEED EXISTS FOR OPTICAL BACKPLANE ENGINEERING DESIGN AUTOMATION

Replacement of Electrical Optical Backplane with Optical Technology

High speed networks for global communication are requiring more bandwidth than the current electrical technology can provide [Grimes, 1997]. This problem has resulted in a need for a replacement technology to eliminate the speed bottleneck, specifically by replacing older "copper-based" electrical optical backplane technology with newer forms of optical backplane technology [Grimes, 2004].

Case Study Focus on FSOI Technology for Automation of Design Risk Mitigation

Although in the field of optical backplane engineering there are three distinct types of technology areas currently available that the AIDF system can support, i.e. fiber-based, light-wave, and Free-Space Optical Interconnect (FSOI), the case study in this thesis focuses primarily on the latter for the purposes of demonstration of the AIDF design environment capability. Although the AIDF system can provide decision support for all three types of systems, the primary focus of the AIDF system demonstration is on the Free-Space Optical Interconnect (FSOI) case, primarily because this area is considered a frontier technology that may benefit the most from an KBE SoS application that enables OBIT designers to make precise design decisions through automated risk-mitigation techniques. Hence, a great impact on the design process can be achieved in the area of FSOI within optical backplane engineering by automation of best practices and domain expertise captured by an AIDF application.

FREE-SPACE OPTICAL INTERCONNECT CASE STUDY OVERVIEW

Primary FSOI Thesis used as Basis for Case Study Implementation

The FSOI application we have examined for automating an actual FSOI is based on the research done by Ayliffe's FSOI group in Montreal, supported by a Grant from the Canadian Institute for Telecommunications Research, as part of the National Center of Excellence program of Canada and NSERC and FCAR postgraduate fellowships – the primary paper we are basing the component decomposition is on this paper [Ayliffe et al., 2001]. Much of this work is further detailed in Ayliffe's PhD dissertation [Ayliffe, 2001]. We referenced this optical backplane thesis here once, which represents much of the data presented in this section for automation. The FSOI design rules are provided by the same group's research that is consistent with this particular implementation with optical backplane components [Kirk, 2003]

Emphasis on Reliability Engineering of Components and Subsystems

FSOI is a good case study to apply design automation within optical backplane engineering for a number of reasons. FSOI systems have remained in a laboratory phase of development for an extended period of time due to many design risk issues, particularly in the domain of opto-mechanical packaging, which is impeding commercialization of the technology [Ayliffe, 1998]. In fact, one of the most important obstacles in the field

of FSOI deals with cost and manufacturability of the optical and opto-mechanical com-

ponents and optimization of cost and volume for packaging techniques. Since the 1990s,

FSOI technology utilizing two-dimensional (2D) optoelectronic device arrays were

thought to be a possible solution to the bandwidth bottlenecks that were expected to soon

appear in computing and telecommunications switching systems [Ayliffe et al, 2001; Ny-

quist et al., 2000]. However, FSOI technology had to overcome some challenges and

introduce systematic design rules [Kirk, 2003] before widespread adoption was to be

taken, primarily in the area of reliability and design risk mitigation. Thus, in order to

demonstrate the effectiveness of the AIDF system on supporting systems engineers with

design reliability, an Artificial Intelligence Design Framework (AIDF) system emphasiz-

ing risk mitigation has been developed.

Modular Assembly of Components

An FSOI is a communications system that uses modulated light to convey infor-

mation. In order for the light to be useful, the direction and intensity of the light, for in-

stance, must be controlled. The characteristics of the light can be controlled through a set

of refractive and defractive elements placed in series with the light as it travels in "free-

space". An implemented FSOI technology can be divided into four modular assemblies

primarily to facilitate assembly and servicing of the optical interconnect [Kirk, 2003; Ay-

liffe et al. , 2001]. All the data concerning the FSOI case study can be traced to Ayliffe's

work in Canada, so we only cite once.

For the modulator-based system shown (Fig 4.1), the optical interconnect is divided into four functional modules with constituent components (Table 4.1):

1. *Beam combination module (BCM)* routes the receiving, read-off, and transmitting beams propagating through via interconnect.

2. *Relay module* propagates the transmitting beams from one stage to the next.

3. *Chip module* performs the electronic-to-optical and optical-to-electronic conversions between the motherboard electronics and the optical interconnect.

4. *Optical power supply (OPS)* generates a 2 – Dimensional (2-D) array of continuous wave (CW) "read-off" beams aligned onto the modulators

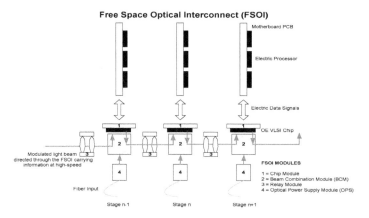

Fig. 4.1. Free-space optical interconnect divided into four modules.

Table 4.1. FSOI divided into four modular subsystems and components.

#	Module	Function	Modular Subsystems and Components
1	*Beam Combination Module (BCM)*	*Routes* the receiving, read-off and transmitting beams through the interconnect	**M1: Beam Combination Module (BCM)** 	Dowel Pins	IP	PMG
IP	PBS	Ruby Ball	
2	*Relay module*	*Propagates* the transmitting beams from one stage to the next	**M2: Relay Module (BCM)** 	Adjustment Mechanism	Precision Ground Rods
Minilens array		
3	*Chip module*	*Performs* the electronic-to-optical and optical-to-electronic conversions between the motherboard electronics and the optical interconnect	**M3: Optoelectronic VLSI Chip Module** | Copper Heatspreader	Minilens holder	Minilens array	Miniprobe thermister
Flex-PCB mount	Minilens spacer	Mounting spacer	OE-VLSI Chip
Four-layer flex PCB	Minilens spacer	Omnidirectional heatsink	Thermoelectric cooler	
4	*Optical power supply*	*Generates* a 2-D array of continuous wave (CW) read-off beams aligned onto the modulators	**M4: Optical Power Supply (OPS) Module** Risley Prisms

KNOWLEDGE ENGINEERING TECHNIQUES FOR PRODUCT ENGINEERING

In the past, knowledge engineering consisted of elicitation of tacit and explicit knowledge from domain experts, usually by trained experts. Today, however, it is possible to have authenticated knowledge development sessions from any point of presence using pre-prepared templates by knowledge engineers (Fig. 4.2). Knowledge bases having very different formats can be developed for knowledge repositories by overlapping responsibilities by multiple domain experts in this approach. Since this knowledge development and updating process can be accomplished asynchronously and concurrently with respect to the design process session, knowledge engineering can be considered an independent function of the automation of the design process. We will show how the AIDF design and inference engine in the KCE block utilize this knowledge later using the current project database for culling and assembling the relevant knowledge being archived into the knowledge repositories. In this section, we give examples of various means that can be used to develop the knowledge base for optical backplane engineering, which can be expanded to include other disciplines seeking reliability engineering automation.

Knowledge Engineering Process
with multiple domain experts developing each knowledge base
in repository for automation by design and inference engine

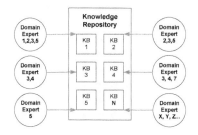

Fig. 4.2. Knowledge engineering process with multiple domain experts for each KB.

Knowledge Engineering with Ontologies

The FSOI modular subsystems and components can be elicited from domain experts in the field. These components can then be conveniently converted into a language that can be processed by intelligent agents on the Semantic Web. Specifically, by using an ontology editor, such as CmapTools, a domain expert can create ontologies that can be easily exported directly into XML, in addition to Web Ontology Language (OWL), as we tested by populating the field with FSOI components (Fig 4.3) and exporting (Fig. 4.4). In this way, we believe any type of system can be hierarchically decomposed using this method into its constituent subsystems and components, and then converted into machine language for processing. This allows for components to be marked up for retrieval by intelligent agents. Furthermore, networked ontologies can be created as new relationships between scattered components can be further expanded. Metrics exist for validating these ontologies in terms of parameters such as breadth, fan-out, and tangledness.

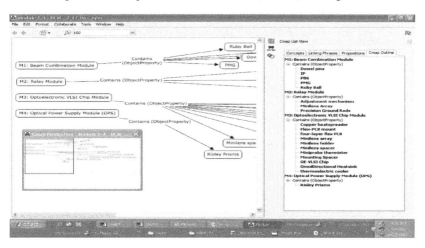

Fig. 4.3. CmapTools mapping concepts for ontologies readable by intelligent agents.

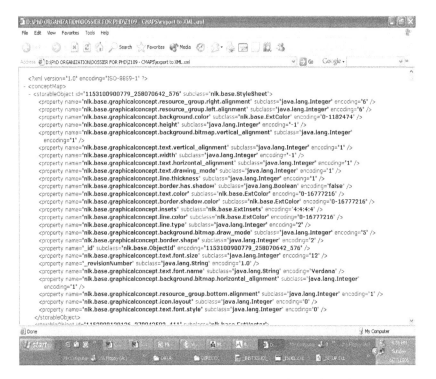

Fig. 4.4. Machine-readable ontologies exported directly from CmapTools.

Knowledge Engineering of Tolerance Analysis

In free-space optical backplane engineering, of the various forms of loss attributed to refraction, diffraction, polarization, or alignment, the alignment makes the greatest impact on efficiency, having four attributes requiring great precision (Table 4.2). We show that alignment can be affected depending on the grouping, or clustering strategy, of the components, e.g. packaging of the Mini-lens with chip results in a variation in the degrees of freedom, in comparison to Mini-lens packaged with optical system. This type of data can be used as input for templates provided on the Web.

Table 4.2 Package module alignment tolerance analysis having high precision.

Degrees of freedom	Mini-lens Packaged with Chip	Mini-lens packaged with optical system
Lateral (x, y)	Micrometers range	Micrometers range
Longitudinal (z)	Micrometer range	Micrometer range
Tilt (X, Y)	1/100 of degree range	1/10 of degree range
Rotational (Z)	1/10 degree range	1/10 of degrees range

Knowledge Engineering of Lateral Alignment

Of the various alignments (lateral, longitudinal, tilt, and rotational) for FSOI, the lateral alignment (Table 4.2) is most critical for tolerance analysis. This knowledge can be captured by having the domain expert provide the accuracy and the limiting factor for each of the various components that impact the lateral alignment, by summing them together until the threshold level is reached. In our case study example, the sum of the above tolerances for the worst case for each leads to a worst case lateral misalignment totaling 80 micrometers, which far exceeds the allowance budget for misalignment significantly at 26 micrometers [Ayliffe, 1998]. This data shows how critical misalignment can be in FSOI applications, so domain expertise on this type of error is important to capture. This type of data can be provided by domain experts via templates on the Web.

Knowledge Engineering for Fault Tree Analysis

The knowledge engineering of a fault-tree analysis can be done with a template resembling a FTA, with places for actual input values (Fig. 4.5) provided on the Web that is filled by a domain expert. This template is in a format that takes into account the level of fault based on the nuclear regulatory commission, provided earlier, ranging from Type

I to Type IV faults. This way the faults can be weighted according to predefined standards, which can be automatically displayed to the domain expert during knowledge engineering before uploading to the central knowledge depository. The AND functions have a multiplier effect and the OR functions have an additive effect to calculate the total probability, e.g. 0.061 or 61 out of 1000 chance of failure of the Chip Module based on these input values by the domain expert. This type of data can be routinely updated and stored by domain experts for future automated calculations and recommendations based on the design and inference results in the engine block.

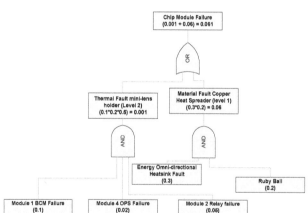

Case study FTA Technique for Calculating Probability of System
Structure Failure and Method for Tracing Problem to Fault Source

Fig. 4.5. Assigning values to FTA based on fault types and interactions.

Knowledge Stored as Axiomatic Rules with Design Rationale

Axiomatic design has two axioms (Table 4.3) from which a set of corollaries (Table 4.4) and theorems are derived, which can form the basis for a conceptual rule-base for axiomatic design.

Table 4.3. Axiomatic design principles for corollary rule-base development.

Axiom	Description	Design Rationale
Axiom 1 (independence axiom)	*maintain the independence of the FR and*	Eliminate all unnecessary coupling for simplification
Axiom 2 (information axiom):	*minimize the information content of the design*	Optimize the number of DP meeting one FR

Table 4.4. Axiomatic corollaries for developing meta-level design rationale.

Corollary	Description	Design Rationale
Corollary 1 (Decoupling of Coupled Design)	Decouple or separate parts or aspects of a solution if FRs are coupled or become interdependent in the proposed design.	Functional independence must be ensured by decoupling if a proposed design couples the functional requirements;
Corollary 2 (Minimization of FRs)	Minimize the number of functional requirements and constraints.	The designer should strive for maximum simplicity in overall design or the utmost simplicity in physical and functional characteristics.
Corollary 3 (Integration of Physical Parts)	Integration design features into a single physical process, device or system when FRs can be independently satisfied in the proposed solution.	Physical integration is not desirable if it results in an increase of information content or in a coupling of functional requirements.
Corollary 4 (Use of Standardization)	Use standardization or interchangeable parts if the use of these parts is consistent with FRs and constraints.	use standard parts, methods, operations and routine, manufacture, and assembly. Special parts should be minimized to decrease cost.

Knowledge stored as Logic Rules with Confidence

Procedural knowledge can be stored as a series of IF/THEN statements with adjustable confidence levels (Table 4.5). This type of knowledge is a basis for automation done by both the inference engine and design engine risk calculations.

Table 4.5. Knowledge stored as logic rules for capturing confidence level.

IF	THEN/DO	A/O	THEN/DO	A/O	THEN/DO	Confidence
IF-1 C1 > 0.002	THEN-1-1 DO Procedure X	AND	THEN-1-2 DO Calculation Y	AND	THEN 1-3 Algorithm Z	45%
IF-2 C2 < 0.001	THEN-2-1 DO Algorithm Z	OR	THEN-2-1 DO Optimization F	N/A	N/A	20%
IF-3 (C1 +C2)/ 4 >> 0.1	THEN-3-1 DO Calculation Y	N/A	N/A	N/A	N/A	35%

Knowledge Engineering with Design Structure Matrix

Each of the FSOI components can be listed in this fashion to capture their interactions (Fig 4.6 a ,b, c) in terms of information exchange, energy exchange, or material exchange on a scale from -10 to 10, where the number indicates positive or negative vulnerability in the case of any given two components interacting. As the number of components increase, the matrix can be expanded and made multidimensional. This format is ideal for machine processing and easy for updating by the domain expert.

Table 4.6. Capturing interactions in terms of (A) information exchange, (B) energy exchange, or (C) material exchange.

(A)	C1	C2	C3
C1		2	3
C2			
C3	-2		

(B)	C1	C2	C3
C1		-7	3
C2	2		
C3		8	

(C)	C1	C2	C3
C1		2	3
C2			
C3	-2		

Knowledge Engineering for Storing Characteristics for Each Interaction

After a DSM has been constructed by the domain expert, the knowledge in terms

of functions, component comparisons, and weighted ratings can be captured for the mod-

ules, as subsystems, and their constituent components (Table 4.7).

Table 4.7. Knowledge that can be integrated from various DSM input templates.

DSM Elements	(M1) Relay Module and (M2) Chip Module Interaction template		
Function of M1	Routes the receiving, read-off and transmitting beams through the interconnect		
Function of M2	Performs the electronic-to-optical and optical-to-electronic conversions between the motherboard electronics and the optical interconnect		
Domain expert Rating	Information (2)	Energy (-7)	Materials (2)

Knowledge Engineering with Technology Risk Factor as a Multiplier Effect

Domain experts can introduce a technology risk factor (TRF) similar to NASA

current model, where the rating for a component varies from high risk to low risk, where

the highest risk components are usually considered to be the untested versions, compared

to those already in use [Brady, 2002]. This can be introduced into a matrix as a multi-

plier effect (Table 4.8). This method is ideal for applying adjustable risk weights to each

of the components.

Table 4.8. Knowledge engineering of technology risk factor multiplier.

A	TRF	C1	C2	C3
C1	3		4	6
C2	2			
C3	7	-14		

Knowledge Engineering of Component Cluster Vulnerability Analysis

Knowledge engineering of Failure mode and Effects Analysis (FMEA) include a description of the incident, the components in direct and indirect contact. This template is ideal for capturing knowledge that relates more than 2 interacting components, while associating their interactions (Table 4.9).

Table 4.9. Knowledge engineering with component vulnerability analysis.

Incident #	Description of Incident	Components in direct interaction	Components in indirect interaction	Level Fault
1	BCM Toxic melt-down	C1, C2, C7	M1, C2, M4	Level I
2	MiniLens holder crack	C3, C5, C9, M2	M1, C1, C2	Level II
3	Contamination of Laser	C2	C5	Level IV

Knowledge Engineering of Case-Based Reasoning

Knowledge can be stored for lateral tolerance, tilt tolerance, maximum array size, optical invariant, including examples, and diagrams as a basis to create a knowledge base for design rationale based on pattern recognition, with respect to design configurations. Here are examples of five different design variations and their associated design parameters storable in the knowledge base (Fig. 4.6) [Kirk, 2003].

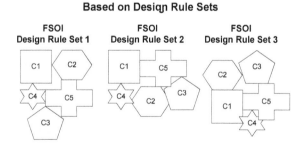

Fig. 4.6. Example FSOI design variations based on design rules

Knowledge Engineering of Reliability over Time-Domain

The availability of components can be provided as input by the domain expert in this format. This is ideal for machine processing and especially useful for component influence analysis conducted by the Conant module in the design engine block. We can gather component histories (heritage) for reliability and determine the relative availability for each component when operating in clusters for three different cases. For i nstance the mirror in Case I is operational 100% of the time when not coupled to the laser, but is o b-served, after coupling in Case II, to begin faulty behavior at time = 3 hours, which co n-tinuous throughout the measured duration of 100 hours, resulting in 2% avai lability. One can conclude that a Type IV catastrophic meltdown might have occurred based on this data, which the domain expert specifies for the clustered components. This data can be used by the CTA module, as the one we conceptualized (Fig. 4.7).

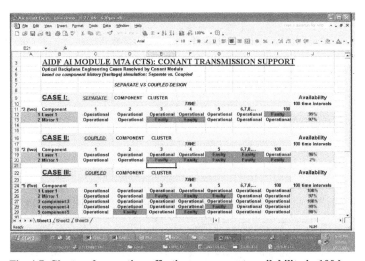

Fig. 4.7. Clustered operation affecting component availability in 100 hrs.

AIDF ARCHITECTURE FRAMEWORK CONFIGURATION

MODEL-DRIVEN ARCHITECTURE SIMPLIFIES CONFIGURATION PROCESS

Assignment of Multiple Design Parameters per Functional Requirement

The AIDF model-driven architecture provides for flexibility in configuration. We have developed an architecture framework model for configuring a KBE SoS application for automated design and inference support for product engineering. In the AIDF implementation developed with the architectural framework development tool Acclaro Design for Six Sigma (DFSS), we show how approximately 300 functional requirements and design parameters are hierarchically decomposed into their constituent elements. The functional requirements (FR), specifying the need, provide a functional specification that can survive many architecture configuration and technology modifications over time, allowing for technology independence and evolution. The design parameters (DP), specifying the modules in the engine block, for instance, provide a means for the architect to assign any number of modules per functional requirement. This way, the architect can adapt to the needs of the end-user by concentration his effort on the configuration of the modules and elements. If needed, he can expand the scope of the AIDF itself, by simply adding more FR to meet the needs of the new end-user and decomposing those needs for further analysis, followed by the same process of assigning appropriate DPs to these FR.

Benefits of AIDF for Concurrent Development and Selection of Best OO Designs

We will analyze the FR mapped to the DP of the reliability module. By focusing our analysis of one particular FR, independent of the other FRs, we demonstrate how to leverage the axiomatic V-model approach for concurrent software development. Thus, after the framework is complete, each of the FR can be outsourced as methods to software engineering teams to develop as OO software modeling, followed by coding. Then all of the FR specified by the AIDF can be re-integrated to produce a KBE SoS. Another benefit is that competing methods can be supplied for one FR, so that the best OO structural model design is selected before coding actually begins.

CASE STUDY RISK MITIGATION NEED

Motivation for Configuring Framework for Reliability

As stated previously, the components of an FSOI require high-precision design support that can benefit from automated design assistance, where design risk mitigation is a chief concern. Since our focus is on reliability engineering for design risk mitigation, we have selected the design of a Free-Space Optical Interconnect (FSOI) for AIDF configuration as an ideal case study that would benefit from automated reliability engineering to assist optical backplane design engineers. In order to demonstrate how the AIDF provides a framework to fulfill the functional requirement (FR) specific to reliability engineering, we will focus specifically on the particular design parameter (DP) that corresponds to the reliability (RBD) module in the engine block during Stage II, in order to show interacting architecture elements that resolve structural and component layout design concerns in the embodiment design phase.

RELIABILITY MODULE DELIVERS RISK MITIGATION

There are 300 FR and DP nodes in the AIDF currently. Of these, there are 40 FR and DP nodes in the engine block, which we will analyze more closely in terms of configuration, optimization, and verification. Since we are addressing reliability engineering for the case study, we will specifically focus on the Reliability (RBD) module and its auxiliary supportive module and element operations.

Selecting the Reliability (RBD) Module for Risk Mitigation in the Engine Block

The goal of the AIDF is to provide a framework for software engineering of this reliability capability by applying a systematic architecture-driven approach to specify a KBE SoS with a high-degree of granularity down to the interacting module element level. We identified twenty modules that can be automated for design support. Of these modules, we selected the reliability module for further analysis. The operations shown in this chapter on this primary module fulfilling the functional requirement for design risk mitigation, and its configured support modules, can be applied for all 300 FR/DP nodes in the design matrix of current AIDF for large-scale software engineering. Thus, to demonstrate the case study focus on reliability engineering of optical backplane components, we have selected the reliability module (RBD) for analysis. Especially since the components of an FSOI operate at the micron level, it is critical that the individual component and grouped cluster vulnerabilities are assessed and their inherent risk mitigated during the automated design process support.

Drilling Down to Element Level

We show the configuration of the reliability module with six other supporting auxiliary modules to address the risk mitigation need of optical backplane engineering. In order to analyze the interactions of the configured reliability and its support modules, we will show varying levels of granularity by hierarchical decomposition of the AIDF, starting from the framework level down to the element level. Higher resolution with increasing granularity is achieved with decomposition in this order: (1) Framework-level, (2) block level (3) engine-level, (4) module-level, and (5) element-level (Fig. 4.8).

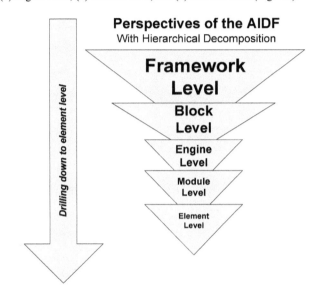

Fig. 4.8. Drilling down to element level.

Configuring the Reliability Module during Stage II

In our case study, most of the automated reliability design support operations on component layout and structure occur during embodiment design phase. Since the design and inference engine are located in the KCE region of the AIDF (block 2), we will concentrate on this aspect of the framework that structures the software development process for engine development. Furthermore, since the case study emphasizes reliability engineering, we will detail the operation of the reliability (RBD) module in more detail in

212

terms of its configuration analysis with respect to other interacting modules and their

elements during AIDF Stage II operation (Fig. 4.9). This particular FR will have

mapped to it multiple modules that support the design process, specifically for design risk

mitigation.

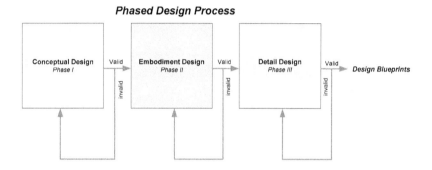

Fig. 4.9. Focusing on the embodiment design stage.

Interactions of the Reliability Module

    In the case of the reliability module fulfilling the design risk mitigation FR, we

are particularly interest in its interaction with the other auxiliary modules supporting its

operation during the embodiment design phase. Thus, from the perspective of modular

interactions with the RBD, we will show how to configure the DSM, TRF, FTA, and

FME, active in the design engine during Stage II, and the DRS and PLS modules active

in the inference engine during Stage II with this primary module to satisfy the need speci-

fied by this particular FR. The configuration process will be done with FR/DP and

DP/DP design matrices, whose interactions will be depicted in detail with component

diagrams.

FRAMEWORK LEVEL CONFIGURATION ANALYSIS

Scoping Configuration Function Analysis at Framework-Level Operations

We focus on the framework to demarcate the block operations (Fig 4.10).

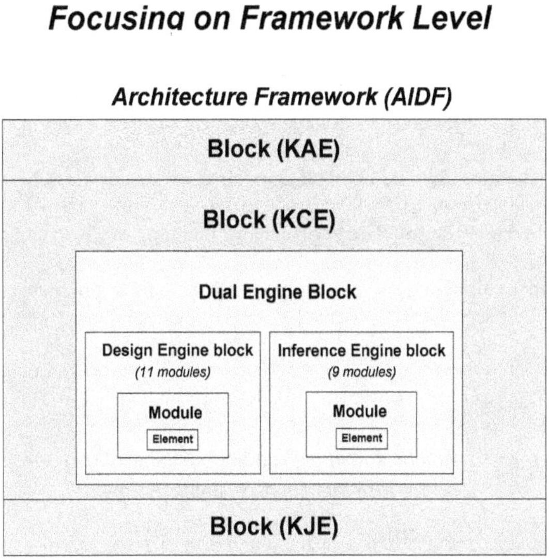

Focusing on Framework Level

Architecture Framework (AIDF)

Fig. 4.10. Focusing on framework level configuration.

Architecture-driven Development for Configuration

Within the scope of design risk mitigation to support the design process, we focus the scope of configuration function analysis of this thesis on developing an architectural framework for reliability engineering (Fig. 4.11). Software engineering, coding, and development are initiated after the framework is delivered to the software and systems engineering to finalize the full-scale KBE SoS development based on the verified and validated specifications provided by the AIDF. This approach especially saves on cost, since implementation is not begun until the framework configuration is completed and the re-

quirements are clearly identified for full-scale production.

The AIDF architectural framework can be represented as an FR/DP design matrix

that maps functional requirements to multiple design parameters, depending on the appli-

cation configuration. Each FR at each node can be further broken down into its constitu-

ent FR and associated DP for detailed module specification. The leaves of the FR even-

tually become the methods defined by in OO modeling languages, i.e. UML. After the

framework is delivered to the software engineers for mapping of FR to OO methods, the

software development modeling and coding process can begin to produce a KBE SoS as

output.

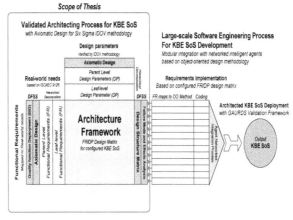

Fig. 4.11. Architecture-driven software engineering for KBE SoS.

BLOCK-LEVEL CONFIGURATION ANALYSIS

Scoping Configuration Function Analysis at Block-Level Operations

We will focus on block level analysis (Fig. 4.12).

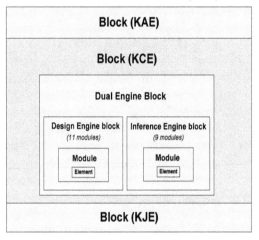

Fig. 4.12. Focusing on block level configuration analysis.

Focusing on EDS/KCE Operation for Reliability Automation Configuration

Within AIDF Stage II, we focus the scope of configuration function analysis on the second processing block, the EDS/KCE, involved in the reliability engineering design calculations and inference mechanisms (Fig. 4.13). Parallel to the other two dynamic stages, this stage is divided into three operational blocks: (1) EDS/KAE block (2) EDS/KCE block, and (3) EDS/KAE block. The first block is involved in knowledge as-similation, the second in reliability engineering, and the third in providing justifiable user output recommendations. In this stage, we will focus only on the operations in the EDS/KCE, since the mechanisms that drive reliability engineering, the design and infer-ence engine blocks, reside here.

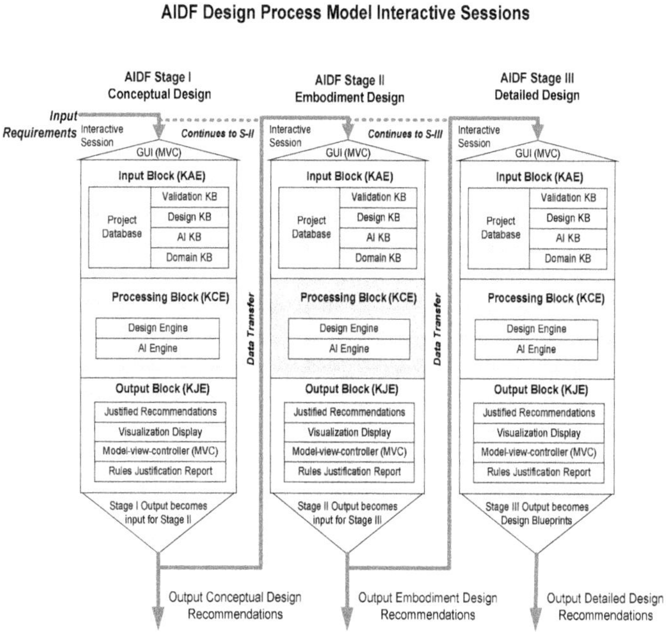

AIDF Design Process Model Interactive Sessions

Fig. 4.13. Focusing on the Processing Block in the EDS/KCE.

The design engine houses the set of modules involved in design support and the inference engine provides houses the set of modules involved in artificial intelligence. Each of these modules are configured before deployment based on the needs of the appli-cation and can be manually enabled by the designer, as part of the reconfiguration proc-ess after deployment, as needed. Later we will show how these two engines can be con-figured by selecting which modules are enabled for a particular instantiation, such as

OBIT design. Then we will show how each of the modules can be connected together at the structural model level. Each of the modules are also connected to the WWW with intelligent agents interacting between the AIDF and the world knowledge repositories, which are being continuously updated though template-based knowledge engineering process, during the design process support.

ENGINE-LEVEL CONFIGURATON ANALYSIS

Scoping Configuration Function Analysis at Engine-Level Operations

Within AIDF the EDS/KCE block, we focus the scope (Fig. 4.14) of configuration function analysis on the AI Design Engine block, housing the design and inference engines involved in the reliability engineering design calculations and inference mechanisms.

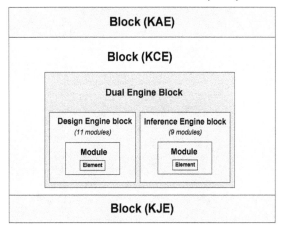

Fig. 4.14. Focusing on dual engine block.

Engine Block Detail

In the AIDF, this central block in Stage II contains a Centralized Knowledge As-similation Engine (CKAU) correlation engine, AI Design Task Manager (ADTM), and AI Design Engine Block (ADEB), shown here again for convenience (Fig 4.15). The CKAU helps coordinate the allocation of knowledge based on the type of task, synthesis or analysis, demanded for support by the ADTM. This architectural component selects the appropriate subtask which activates and appropriate engine block, which, in turn, ac-tivates the particular set of modules configured for the instantiation, e.g. case study on optical backplane engineering. Once activated, the module executes the algorithm in-structions and repeats as necessary.

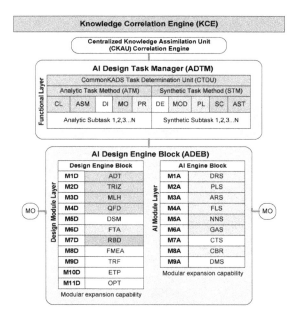

Fig. 4.15. Knowledge correlation engine revisited before configuration.

Dual Engine Block Operation

The dual engine block can operate in tandem after connecting, or interlacing, the modules together (Fig 4.16). Later, we will show how to interlace the engine block for configuring the OBIT application for intra-module and inter-module operations within the same engine block and between the engine blocks using the framework (Fig. 4.17). Furthermore, we will show how to interlace the module elements, specified by the configured AIDF, for interaction within modules, as well as interaction between modules, for specifying intra-module and inter-module operation, respectively. These interactions represented by the framework-level, engine-level, module-level, and element-level can later be used to develop UML that meets the highly specified framework's requirements, as in the case specified for reliability engineering in the engine block.

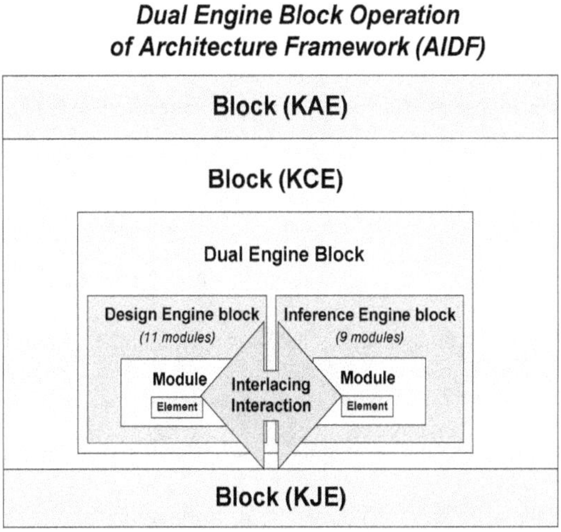

Dual Engine Block Operation of Architecture Framework (AIDF)

Block (KAE)

Block (KCE)

Dual Engine Block

Design Engine block (11 modules)	Inference Engine block (9 modules)
Module Element	Interlacing Interaction	**Module** Element

Block (KJE)

Fig. 4.16. Dual engine operation of AIDF with interlacing connections.

**Interlacing Connections between Dual Engine Block
for Reliability Automation of Configured Case Study**

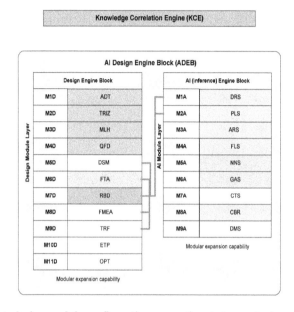

Fig. 4.17. Interlacing module configuration connections between dual engine block.

CASE STUDY OPERATION OVERVIEW OF EDS BLOCK ENGINE

In this section, we describe operation of the EDS block engine during Stage II, based on the detail provided in the AIDF chapter for the engine block operation. We show a diagram of the operation of the AIDF with external entities during Stage II (Fig. 4.18). The larger diagram at the beginning of Chapter III provides a more detailed view of how Web Services are actually activated, by focusing on the reliability engineering module, i.e. RBD. This diagrams shows how all the KBE SoS elements are related in their dynamic operation.

EDS/KAE Block 1

During EDS/KAE for the case study, block 1 is involved in assimilating quantitative knowledge on optical backplane components are collected, gathered, and combined from the user and an environment (data flow from AIDF and Web Services). Many of the knowledge engineering techniques captured using template formats, described earlier in the knowledge engineering section, such as DSM and marked up optical backplane components in XML and OWL are retrieved by intelligent agents and assembled in the KAE blackboard in preparation for processing in the EDS/KCE (Block 2).

EDS/KCE Block 2

During EDS/KCE Block 2 for the case study, block 2 is involved in coordinating, tasking, and processing the influx of knowledge, in the form of updated rules, if any, and other marked-up components in OWL format from the KAE. Since reliability engineering is considered an analysis task, analysis subtasks are executed. The task manager selects the appropriate subtask, i.e. design/configuration, which executes the appropriate module, i.e. Reliability (RBD), which, in turn, executes the OO elements defined in the module by software engineers. Some of the elements may invoke intelligent agents and other interlaced modules, depending on the configuration and internal UML structural models (outside the scope of the thesis).

In this block, reliability engineering tasks for optical backplane engineering are conducted at the component and structural level, such as design structure matrix (DSM) for component-component interaction, failure-mode and effects analysis (FMEA) for vulnerability assessment, and reliability block diagram (RBD) analysis for calculating series and parallel reliability, and fault tree analysis (FTA) to pinpoint faults. Domain rule support (DRS) and predicate logic support (PLS) are provided as supportive inference mechanisms invoked by the primary module, i.e. RBD.

EDS/KJE Block 3

During EDS/KCE for the case study, block 3 is involved in justifying the automated *quantitative* recommendations for risk mitigation done by providing an interaction mechanism using an MVC, justifiable recommendations, and a rich visualization display.

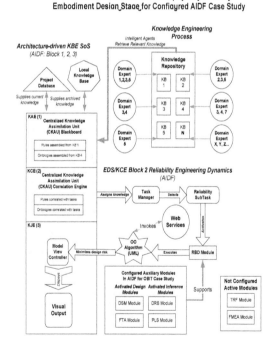

Fig. 4.18. Reliability engineering dynamics defined by AIDF EDS/KCE.

MODULE-LEVEL CONFIGURATION ANALYSIS

Scoping Configuration Function Analysis at Module-Level Operations

In order to show the interacting modules, we focus on the module level within the framework (Fig 4.19).

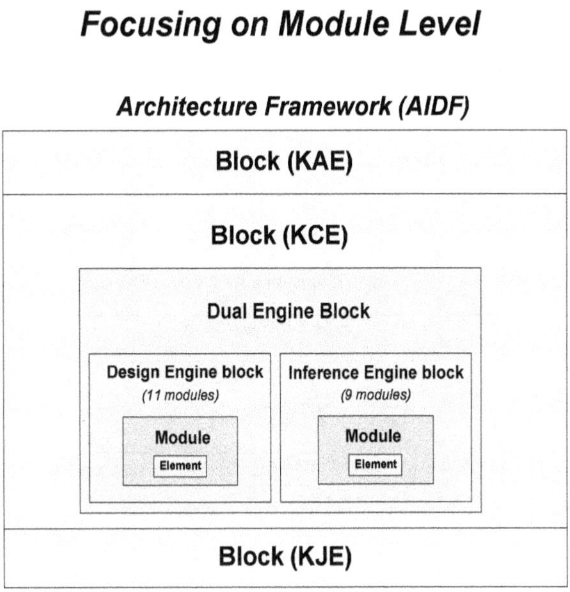

Focusing on Module Level

Architecture Framework (AIDF)

Fig. 4.19. Focusing on module-level operations.

Designation of Design Matrix Configuration for Active Modules

Within AI Design Engine Block (ADEB), we focus the scope of configuration function analysis on the design and engine block, *individually,* showing the framework design matrix before (Table 4.12, 4.13) and after (Table 4.14, 4.15) configuration for both the design and inference engine modules in their respective engine blocks. The only engines active in the design engine block during stage II are DSM, FTA, RBD, and TRF, and in the inference engine block are DRS and PLS (Table 4.11). For configuration of the OBIT KBE SoS, we show which modules (as DPs) fulfill the needs of a FR for the engines providing automated reliability engineering design support in real-time for the KAE in Stage II block 2 with "X" indicating pre-configuration expected module operation, "C" indicating configured framework for modular operation, and "N" indicating an unneeded module for that particular functional requirement, which is remains not configured (Table 4.10).

Table 4.10. Identification of preconfigured, configured, and active modules.

Code	Description KCE (Stage II/Block 2)
X	Active Module in KCE (Stage II/Block 2)
C	Configured Active Module in KCE (Stage II/Block 2)
N	Not configured Active Module in KCE (Stage II/Block 2)
X	Active Configured Module
I	Interacting Active Configured Module
R	Interacting Active Configured Module with Reliability (RBD) module

Primary and Auxiliary Modules

For instance, in the case of the configured OBIT for the design engine block, the marked RBD module is the primary active module with two other auxiliary modules providing necessary support. The module marked N is not configured for this particular functional requirement. Hence, multiple modules represented as DPs can fulfill the needs of one FR, traceable to end-user needs to assist in verification process of predeployment DFSS IDOV front-end validation. Later we will show how to optimize the framework by reducing the number of DPs for achieving the same FR using the IDOV method. Later, we will also show how to configure the AIDF for dual engine operation having interac t-ing design and inference engine modules for an application, after each of the engine blocks are configured alone.

The dual configuration is optional and at the discretion of the end-user, where each engine can be configured independently or together, if necessary, for interacting e n-gine operation. This decoupled configuration feature is attributed to the inherent modula r-ity of the AIDF allowing for this flexibility of decoupled configuration, if needed. One can hierarchically decompose the design and inference engine block starting with the top -level module identified, where each FR can be further broken down into hundreds of co n-stituent elements. This process of axiomatic design ensures that each need is accounted for in each module, and allows for connecting between elements between sub -elements of each module for detailed configuration.

Table 4.11. EDS/KCE active tasks/design and inference modules.

| EDS/KCE (S2/Block 2) | Correlation of structural ele-ments and constraints with appropriate support engines in a way so that the designer's activity is supported by CommonKADS framework for KBE | Primary Active Task: Analytic Tasks, Syn-thetic Tasks

Primary Active Design Modules:

DSM, TRF, FTA, RBD, FMEA

Primary Active AI Modules:

DRS, PLS |

Table 4.12. Stage II pre-configuration design engine modules FR/DP design matrix.

		DP1	DP2	DP3	DP4	DP5	DP6	DP7	DP8	DP9	DP10	DP11
FR1	ADT											
FR2	TRIZ											
FR3	MLH											
FR4	QFD											
FR5	*DSM*					X						
FR6	*FTA*						X					
FR7	*RBD*							X				
FR8	FMEA											
FR9	*TRF*									X		
FR10	ETP											
FR11	OPT											

Table 4.13. Stage II pre-configuration inference engine modules FR/DP design matrix.

		DP1	DP2	DP3	DP4	DP5	DP6	DP7	DP8	DP9
FR1	*DRS*	X								
FR2	*PLS*		X							
FR3	ARS									
FR4	FLS									
FR5	NNS									
FR6	GAS									
FR7	CTS									
FR8	CBS									
FR9	DMS									

Table 4.14. Stage II configured OBIT design engine modules FR/DP design matrix.

		DP5	DP6	DP7	DP8	DP9
FR1	ADT					
FR2	TRIZ					
FR3	MLH					
FR4	QFD					
FR5	*DSM*	X	C	C	N	N
FR6	*FTA*	N	X	N	N	N
FR7	*RBD*	C	C	X	N	N
FR8	FMEA	N	N	N	X	N
FR9	*TRF*	N	C	N	N	X
FR10	ETP					
FR11	OPT					

Table 4.15. Stage II configured OBIT inference engine modules FR/DP design matrix.

		DP1	DP2	DP3	DP4	DP5	DP6	DP7	DP8	DP9
FR1	*DRS*	X	N							
FR2	*PLS*	C	X							
FR3	ARS									
FR4	FLS									
FR5	NNS									
FR6	GAS									
FR7	CTS									
FR8	CBS									
FR9	DMS									

Configuration Function Analysis for Dual Engine Interacting Block Operations

Within AI Design Engine Block (ADEB), we focus the scope of configuration function analysis on the dual engine block, where the design and engine block are operating in *combination* (Table 4.16). The configuration now can interconnect the operations between each engine block, so that the design engine can get auxiliary support from the inference engine modules, and vice versa, as needed, to fulfill any given FR. The support is not necessarily symmetrical, i.e. the design engine modules may be configured for more support from inference engine modules, than vice versa, for instance. Any number of modules can be configured so that they are available among the active modules executed during each stage.

Table 4.16. Stage II configured OBIT dual engine block.

#	FR	Design/Inference	1	2	3	4	5	6	7	8	9	10	11	12	13
		Modules (DP)	Interacting Design and Inference Modules in Configured Dual Engine Block												
1	FR1	ADT													
2	FR2	TRIZ													
3	FR3	MLH													
4	FR4	QFD													
5	FR5	DSM					X	C	C	N	N			N	N
6	FR6	FTA					N	X	N	N	N			C	C
7	FR7	RBD					C	C	X	N	N			C	C
8	FR8	FMEA					N	N	N	X	N			N	N
9	FR9	TRF					N	C	N	N	X			C	N
10	FR10	ETP													
11	FR11	OPT													
12	FR1	DRS					N	C	N	N	C			X	N
13	FR2	PLS					N	N	C	C	N			C	X
14	FR3	ARS													
15	FR4	FLS													
16	FR5	NNS													
17	FR6	GAS													
18	FR7	CTS													
19	FR8	CBS													
20	FR9	DMS													

Scoping Configuration Function Analysis for Reliability Functional Requirement

Within AI Design Engine Block (ADEB), we focus the scope of configuration function analysis on just one of the FR, fulfilled by the reliability (RBD) module where the design and engine block are operating in *combination* (Table 4.17). Since we are primarily concerned with reliability engineering in this thesis, we will select module 7, the reliability (RBD) module for further analysis. Thus, we know that this particular FR is fulfilled with one primary RBD module with two auxiliary design modules and two

auxiliary inference modules providing support to achieve the FR robustly, instead of relying on just one module. In this way, each of the FR can be sliced and outsourced to a
software engineering development team. Later, once all of the software engineering
teams have developed the FR, with respect to the multiple DP characterization, the
framework integrates all of the hundreds of FRs into a one unified, cohesive, large-scale
software implementation of a KBE SoS. There can be multiple solutions for each FR, so
that the best solution in UML is selected.

Table 4.17. Scoping of FR for slicing and sending to software development team.

#	FR	DP Select Design Module	Interacting Design and Inference Modules in Configured Dual Engine Block												
			1	2	3	4	5	6	7	8	9	10	11	12	13
7	FR7	RBD					C	C	X	N	N			C	C

MODULAR INTERACTION ANALYSIS OF DUAL ENGINE BLOCK

Component Diagram Showing Feed-Forward Data Interchange

Once we have established the FR and its primary and auxiliary modules, then we
can concentrate on refining the type of interactions between the modules. Here is the
configuration defining the connection between the modules in the engine block for the
primary RBD module (Module 7-X) (Fig 4.20). The RBD module can be represented as a
design structure matrix, showing the interacting components, i.e. modules. Thus Modules 7-X is feeding data to Module 5-C and Module 13-C, while receiving data from
Module 6-C (FTA).

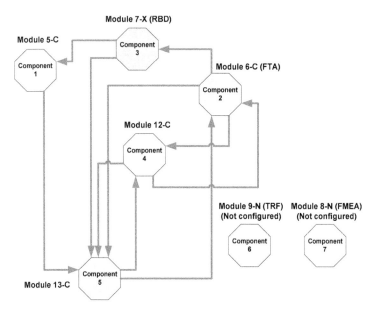

Fig. 4.20. Dual Engine modules interaction diagram.

DSM Analysis of Feed-Forward Modules

In order to further define the type of interactions occurring (Fig. 4.17), we can construct a DSM of the module interaction. We define the interactions occurring between the configured modules as "I" and the interacting configured modules interacting with RBD as "R" (Table 4.18). These interactions represent feed-forward data exchange. We are interested in further definition of the RBD, so we note that the M7X (RBD) is being fed data by M6C (FTA), while M7X feeds data to M5C (DSM) and M13C (PLS).

Table 4.18. Component DSM analysis with five configured feed-forward modules.

		C1	C2	C3	C4	C5
		M5C	M6C	M7X (RBD	M12C	M13C
C1	M5C	X				I
C2	M6C		X	R	I	I
C3	M7X (RBD)	R		X		R
C4	M12C		I		X	I
C5	M13C		I		I	X

Modular Data Transfer Rate for Design Output and Justified Rationale

The data rate transfer for the RBD interaction can be defined here in milliseconds and also represented as a DSM (Table 4.19) and diagram (Fig. 4.21). Hence, the frame-work model shows the module connections and operations speed between engine mod-ules and their elements, as well as elements within modules. For instance, the data trans-fer rate between M6C and M7X is 20 ms, as M6C feeds data to M7X processed by inter-nal mechanism modeled by UML diagrams defining the operations internally. M7X (RBD) feeds data to M5X (FTA) at a rate of 30ms and simultaneously to M13C at 70ms (PLS). In the OBIT case, the output of the reliability operations in the automated reliabil-ity block diagram module, later modeled internally by UML structural models in detail, provide data to the automated Fault Tree Analysis operations for vulnerability analysis of OBIT components, while sending data to the Predicate Logic Support module for heuris-tic analysis for output of recommendations and design rationale.

Table 4.19. Data transfer rate between modules interacting with RBD.

		C1	C2	C3	C4	C5
		M5C	M6C	M7X (RBD)	M12C	M13C
C2	M6C		X	20ms	1	1
C3	M7X (RBD)	30ms		X		70ms

Data Transfer Rate for OBIT Configuration of Interacting Engine Modules (Components)
(in milliseconds)

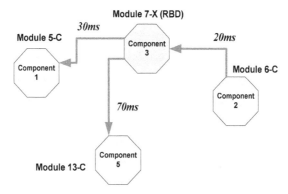

Fig. 4.21. Data transfer rate defined between interacting engine modules.

ELEMENT LEVEL CONFIGURATION ANALYSIS

We will now focus on element level configuration analysis after further hierarchical decomposition of the modules, in order to define the interacting elements (Fig 4.22). The next step of resolution is accomplished by software engineering of the module element itself using OO UML methods, outside the scope of the thesis.

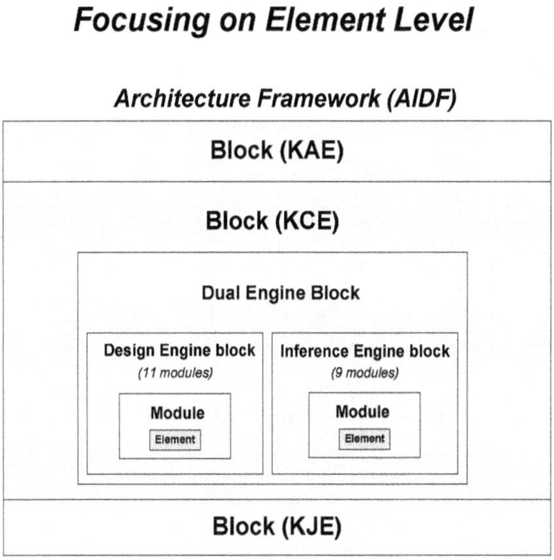

Focusing on Element Level

Architecture Framework (AIDF)

Block (KAE)

Block (KCE)

Dual Engine Block

Design Engine block *(11 modules)*	Inference Engine block *(9 modules)*
Module Element	**Module** Element

Block (KJE)

Fig. 4.22. Element level configuration analysis.

Further Decomposition of Interacting Modules for Operational Granularity

The interacting and configured modules can be further decomposed into elements. In the example of the reliability (RBD) module interacting with three other modules, we can more explicitly define the internal and intra-module operations further by expanding the interactions analysis provided by hierarchically decomposing the module further and analyzing with a DSM again. For instance, OBIT configured module M6C, interacting with Module M7X by feeding data at a transfer rate of 20ms, can be further decomposed (Table 4.21) into three DP elements (M6C1, M6C2, M6C3). The same way, M7X can be decomposed into its two constituent elements (M7X1, M7X2), and the feedforward interaction between the M6C and M7X module can be defined, in addition to the internal interactions of the M7X module independent of other modules. Then further interaction analysis can be done, for instance, between module elements, indicated with an "E", and operations occurring within modules, indicated with a "W" (Table 4.20).

Table 4.20. Identification of interaction type for decomposed engine.

Code	Description of interaction type for decomposed engine
R	Interacting reliability modules having feedforward coupling with RBD
E	Module elements interacting between reliability modules
W	Module elements interacting within reliability modules

Table 4.21. Decomposition of modules interacting with RBD.

			C1	C2	C3			C4	C5
			M5C	M6C	M7X (RBD			M12C	M13C
						M7X1	M7X2		
C2	M6C			X	20 ms			I	I
		M6C1				E	N		
		M6C2				E	N		
		M6C3				N	E		
C3	M7X (RBD)		30 ms		X				70 ms
		M7X1				X	N		
		M7X2				W	X		

Interacting elements analysis for Inter-module and intra-module connection

The AIDF modules and elements can be configured to interact within the same

engine or between engines (Fig. 4.23), as well as inter-module and intra-module, respec-

tively. Furthermore, the module elements can interact. Keeping in mind the scope of the

AIDF having hundreds of FR for any given application, there can be multiple DP (mod-

ules) mapped to each FR, where each configured module and its constituent elements can

be sent to software engineers for detailed structural modeling in UML, followed by

downstream coding. This approach saves on unnecessary iteration cost, by establishing

the requirements early on before any coding begins. Thus, the AIDF provides a high-

level framework that can be hierarchically decomposed, down to the interacting module

element level, for KBE SoS application development based on architecture-driven soft-

ware engineering with a high degree of granularity and resolution.

**Interacting Module Element Configuration Analysis for
Hierarchically Decomposed AIDF Functional Requirements**
(Inter-module & Intra-module)

Reliability Module (7-X) receiving data from Module (6-C) at a rate of 20 milli-seconds

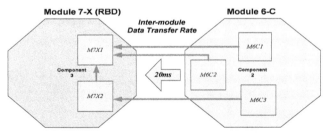

Keeping in mind the scope of the AIDF having hundreds of FR for any given application, there can be multiple DP (modules) mapped to each FR, where each configured module and its constituent elements can be sent to software engineers for detailed structural modeling in UML, followed by downstream coding. This approach saves on unnecessary iteration cost, by establishing the requirements early on before any coding begins. Thus, the AIDF provides a high-level framework that can be hierarchically decomposed, down to the interacting module element level, for KBE SoS development based on architecture-driven software engineering with a high degree of granularity and resolution.

Fig. 4.23. Inter-module and Intra-module interaction analysis.

Integration Process of Axiomatic Architecture-Driven Software Engineering

Once all of these interactions are defined, software engineers can develop UML models to further define the elements within the modules during structural model analysis that can be used by programmers downstream to code the KBE SoS (Fig. 4.24). This process is described in the overall axiomatic V-model used as the process model for architecture-driven software engineering. Once all coding is complete, the framework provides for seamless integration of all modules operating together. For convenience, we provide the axiomatic V-model approach again for AIDF construction, whose functional requirements specifications from high-level to leaf-level FR specifications that are mapped to multiple DP (modules) having interacting elements, can be used as a foundation for architecture-driven software development for a KBE SoS.

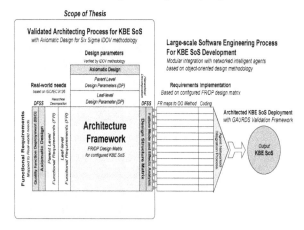

Fig. 4.24. AIDF integration of FR with configured DP for case study KBE SoS.

AIDF ARCHITECTURE FRAMEWORK IMPLEMENTATION

ACCLARO DESIGN FOR SIX SIGMA FRAMEWORK IMPLEMENTATION

We implemented the AIDF architectural framework model (Fig. 4.25), supporting

three design stages, using Acclaro DFSS tool.   In this chapter, our goal is to show how

the framework helps software engineers tackle a large-scale software engineering prob-

lem by developing a design matrix, focusing on reliability engineering for our case study

application on product design using Optical Backplane Interconnect Technology (OBIT).

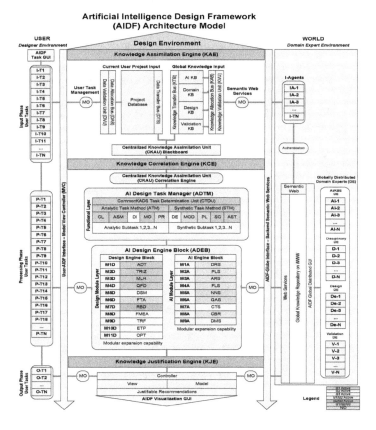

Fig. 4.25. AIDF Architectural framework model detail to be implemented.

Generic Architecture Framework Model Implementation (Pre-Configuration)

We were able to reconstruct the AIDF architectural framework as an implementation in Acclaro DFSS (Fig. 4.26), which allows for hierarchical decomposition of all the functional requirements (FR) and design parameters (DP). Each of the three blocks, the KAE, KCE, and the KJE, for each of the three stages in the AIDF, was entered into the architectural framework development tool Acclaro DFSS for analysis. The architectural style we used in the model was pipes and filters, which fit neatly with the tool implementation of 300 FR and DP nodes (Fig), where the filters were represented as DPs (The "+" mark indicates where they can be expanded for detail viewing).

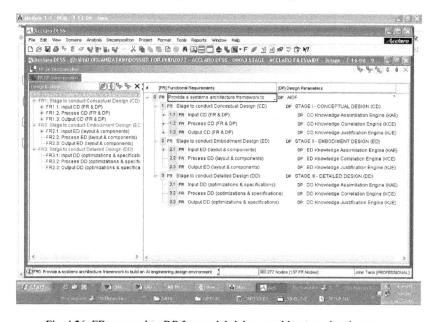

Fig. 4.26. FR mapped to DP for model-driven architecture development.

Hierarchical Decomposition of Framework Maps to Real-World Needs

Currently, the AIDF architectural framework model for the case study consists of

300 scalable Functional Requirements (FR) and Design Parameters (DP) identified as

nodes total. Each FR node has a corresponding DP node in the generic architectural

framework. A tree model view can be selected to see how these FR and DPs are related

to each other within the framework (Fig. 4.27).

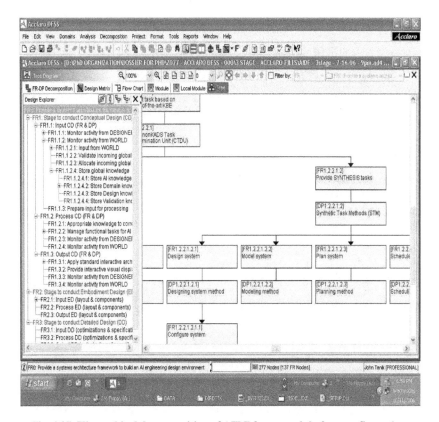

Fig. 4.27. Hierarchical decomposition of AIDF framework before configuration.

Scalability of Functional Requirements and Design Parameters

Depending on the granularity needed, the architectural framework can be scaled according to the needs of the configured application. For instance, an application in the field of mechanical engineering may emphasize a need for robust reliability calculations for series and parallel components. In that case the FR node that corresponds to one of the DP nodes in the engine block, such as the RBD module, can be further defined for that particular case study instantiation. The generic case of the architectural framework before configuration appears with all the "X" marks identified along the diagonal, showing where each DP fulfills a given FR (Fig. 4.28).

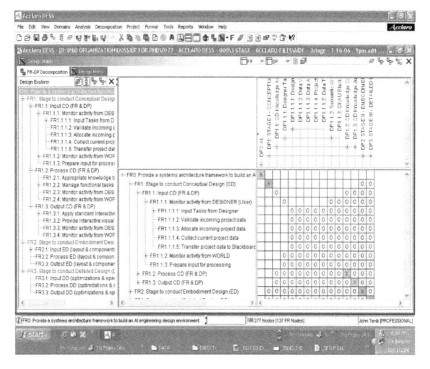

Fig. 4.28. Scalability of the AIDF implementation.

Leaf-level Hierarchy Execution Priority

The flow chart view shows the order in which the modules are executed, according to the leaf-level hierarchy (Fig. 4.29). This is essential for software engineers to know when they receive the architectural framework FR and DP specifications for consistent structural modeling in UML according to the execution priority. This view is useful when the functional requirements are actually sent for coding, since it provides the system analysts an overall view of the intended functional requirements that relates to their work, impacting their decisions on integrating with other teams.

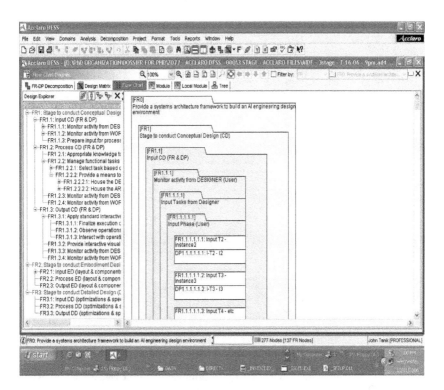

Fig. 4.29. Leaf-level hierarchy execution priority detail.

Systems Engineering Project Management using Acclaro DFSS

The project view shows how the AIDF implementation can be exported into a view convenient for project managers to develop a project management plan for each FR using common application tools such as MS project (Fig 4.30). This type of high-level FR decomposition is very useful for assuring that all the resources available for development of a particular AIDF KBE SoS instantiation are provided on time during software engineering of the configuration based on the AIDF implementation guidelines. This approach can be applied in the case of the OBIT KBE SoS application.

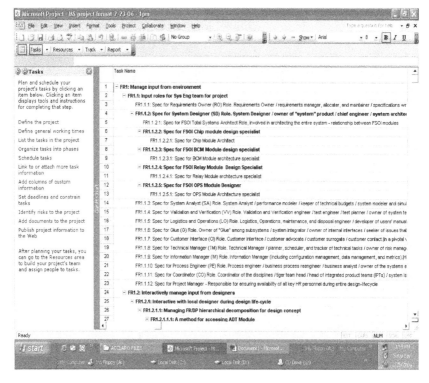

Fig. 4.30. Providing management with decomposed time-line consistent with AIDF.

243

AIDF ARCHITECTURE FRAMEWORK OPTIMIZATION/VERIFICATION

ASSURING QUALITY WITH DESIGN FOR SIX SIGMA VALIDATION

Design for Six Sigma IDOV Methodology Applied to Case Study

In order to validate our architecture framework we apply the DFSS IDOV methodology for the OBIT case study (Table. 4.22). The IDOV methodology is based on the ideas of identify, design, optimize, and verify. This entails showing the relevance of the AIDF architectural framework implementation in meeting the needs of software engineering of a structural model that can be eventually coded to develop an instantiation of a KBE SoS that is configured specifically for optical backplane engineering.

We will show how DFSS IDOV is applied to the case by showing the following:

(1) How we improve architecture by ensuring that functional requirements actually meet the original *end-user* needs by *identifying* real-world needs using method such as Quality Function Deployment (QFD) based on ISO/IEC 9126 defining standard features for end-user;

(2) How we improve architecture by ensuring that proper risk mitigation and reliability engineering methods have been applied in the *design* of the relevant modules of the case study engines;

(3) How we improve architecture by ensuring that efficiency is introduced into design by *optimizing* with DSM and TRIZ;

(4) How we improve the architecture by ensuring that the architected solution actually meets the original *engineering* specifications by *verifying* that all the stated functional requirements are actually being met explicitly by the design parameters of the system by applying axiomatic hierarchical decomposition.

Table 4.22. Design for six sigma IDOV methodology for front-end validation.

Step	Purpose	Method	AIDF case study
Identify	Improve architecture by ensuring that functional requirements actually meet the original *end-user* needs	Identify real-world needs using method such as Quality Function Deployment (QFD) based on ISO/IEC 9126 defining standard features for end-user	IDOV- QFD with ISO/IEC 9126 applied to AIDF for optical backplane reliability engineering
Design	Improve architecture by ensuring that proper risk mitigation and reliability engineering methods have been applied in the modules	Design for systems level and component level risk mitigation by automating only industry-accepted techniques like probability risk assessment (PRA) with Fault-Tree Analysis (FTA) and Reliability Block Diagram (RBD), for reliability engineering using methods such as axiomatic design theory (ADT), Failure Mode and Effects Analysis (FMEA)	IDOV- risk mitigation algorithms in the engine block modules
Optimize	Improve architecture by ensuring that efficiency is introduced into design	Optimize the configuration to ensure that dependencies are resolve using method such as Design Structure Matrix (DSM) and Theory of Inventive Problem Solving (TRIZ)	IDOV-DSM analysis of framework with TRIZ
Verify	Improve architecture by ensuring that the architected solution actually meets the original *engineering* specifications	Verify that all the stated functional requirements are explicitly satisfied by the system DPusing methods such as ADT	IDOV- show that all FR map to at least 1 DP in the framework for case study

Avoiding Requirements Trap with Acclaro DFSS

Various front-end validation techniques are applied to the AIDF to prevent *implementation trap*, a condition where original functional requirements are not addressed or lost during downstream design decisions made during software engineering implementation. We were able to prevent implementation trap by assuring that every functional requirement (FR) defined had a corresponding design parameter (DP) meeting that specification. Furthermore, whenever a FR was expanded, this created more lower-level FRs that needed addressing. The implementation tool's axiomatic decomposition made it convenient to identify which DP was missing. Thus, we avoided getting lost during requirements analysis. This type of analysis is useful for cost-saving down-the-line, as it is much easier to correct earlier than after programming begins.

Avoiding Requirements Creep with Acclaro DFSS

Furthermore, *requirements creep,* a condition of late-stage functional add-ons defining KBE system needs, can not only be avoided, but flexibly addressed by the AIDF due to its modular configurability for each domain of application. This front-end approach is especially useful for validation to meet user expectations by addressing functional requirements that are traceable directly to top-level domain-specific customer needs to avoid requirements creep. We were able to accommodate new requirements as they appeared during our research on an architectural framework that met the needs of the case study using the implementation tool's ability to manage requirements and their hierarchical relationships, adding and deleting as necessary without incident.

Verification with IDOV DFSS

Once all the DPs have been assigned, the hierarchical decomposition of FR provides for complete traceability of the software implementation to high-level functional
requirements, which is crucial for ensuring that the final verified product meets the needs
of the end-user. This process instills quality throughout the case study development
process for a particular configuration by assuring that the generic framework already follows axiomatic principles of hierarchical decomposition. Since each FR maps to one OO
method, approximately half of the 300 nodes are needed to build the current AIDF. Each
of these methods forms high-level requirement specifications for a software engineering
team who construct UML diagrams that meet the needs of the method, and thus the FR.
Once these OO diagrams are complete and verified by the FR traced to the source, the
actual end-user need, the coding process can begin at the programmer level.

Optimization of the Design Matrix during DFSS IDOV

After configuration (Table 4.23 and Table 4.24), there may be more than one DP
assigned to each FR. At this point dependencies are revealed and any unnecessary interactions can be removed by applying the DSM and TRIZ optimization step of the DFSS
IDOV process.

Table 4.23. Stage II configured OBIT dual engine block.

#	FR	DP Modules Design/Inference	Design Parameters for Dual Engine Block Interacting Design and Inference Modules												
			1	2	3	4	5	6	7	8	9	10	11	12	13
5	FR5	DSM					X	C	C	N	N			N	N
6	FR6	FTA					N	X	N	N	N			C	C
7	FR7	RBD					C	C	X	N	N			C	C
8	FR8	FMEA					N	N	N	X	N			N	N
9	FR9	TRF					N	C	N	N	X			C	N
10	FR10	ETP													
11	FR11	OPT													
12	FR1	DRS					N	C	N	N	C			X	N
13	FR2	PLS					N	N	C	C	N			C	X

Of the twenty modules, we will examine only modules 5-13 for optimization purposes (Table 4.24). For optimization of the engine operation defined by the AIDF architecture framework configured for OBIT, our goal is to eliminate as many unnecessary dependencies as possible to achieve the same function. For instance, by using TRIZ indicated by "TRIZ", we can systematically apply innovation to a FR defined by FR7. For example, we can eliminate the auxiliary support module provided by Module 5 and Module 12 by discovering how to implement the RBD module with more internal elements, as opposed to more module linkages (Tables 4.25 and Table 4.26).

Table 4.24. Stage II configured OBIT dual engine block.

#	FR	DP Modules Design/Inference	Design Parameters for Dual Engine Block Interacting Design and Inference Modules												
			1	2	3	4	5	6	7	8	9	10	11	12	13
5	FR5	DSM					X	C	C	N	N			N	N
6	FR6	FTA					N	X	N	N	N			C	C
7	FR7	RBD					C	C	X	N	N			C	C
8	FR8	FMEA					N	N	N	X	N			N	N
9	FR9	TRF					N	C	N	N	X			C	N
10	FR10	ETP													
11	FR11	OPT													
12	FR1	DRS					N	C	N	N	C			X	N
13	FR2	PLS					N	N	C	C	N			C	X

Table 4.25. Stage II DFSS IDOV optimization process before TRIZ.

#	FR	DP Modules Design/Inference	Design Parameters for Dual Engine Block Interacting Design and Inference Modules												
			1	2	3	4	5	6	7	8	9	10	11	12	13
7	FR7	RBD					C	C	X	N	N			C	C

Table 4.26. Stage II DFSS IDOV optimization process after TRIZ.

#	FR	DP Modules Design/Inference	Design Parameters for Dual Engine Block Interacting Design and Inference Modules												
			1	2	3	4	5	6	7	8	9	10	11	12	13
7	FR7	RBD					TRIZ	C	X	N	N			TRIZ	C

Thus, we have now only modules 6, 7, and 12 providing the same functionality as before, except with two less modules. The same process (Table 4.27) can be applied to all the other functional requirements in the design matrix (Table 4.28) indicated by "$". Elimination of entire modules is expected to tremendously reduce cost and speed up operations, since elements within modules achieving the same functionality usually have a faster data transfer rate and less expensive embedded architectural elements, as opposed to modules.

Table 4.27. Stage II DFSS IDOV optimization process with TRIZ.

		DP Modules	**Design Parameters for Dual Engine Block Interacting Design and Inference Modules**												
#	**FR**	**Design/Inference**	**1**	**2**	**3**	**4**	**5**	**6**	**7**	**8**	**9**	**10**	**11**	**12**	**13**
5	**FR5**	DSM					**X**	**C**	**C**	**N**	**N**			**N**	**N**
6	**FR6**	FTA					**N**	**X**	**N**	**N**	**N**			**C**	**C**
7	**FR7**	RBD					**TRIZ**	**C**	**X**	**N**	**N**			**TRIZ**	**C**
8	**FR8**	FMEA					**N**	**N**	**N**	**X**	**N**			**N**	**N**
9	**FR9**	TRF					**N**	**C**	**N**	**N**	**X**			**C**	**N**
10	**FR10**	ETP													
11	**FR11**	OPT													
12	**FR1**	DRS					**N**	**C**	**N**	**N**	**TRIZ**			**X**	**N**
13	**FR2**	PLS					**N**	**N**	**C**	**C**	**N**			**TRIZ**	**X**

Table 4.28. Stage II Design Matrix with less DP per FR after TRIZ.

#	FR	DP Modules Design/Inference	Design Parameters for Dual Engine Block Interacting Design and Inference Modules												
			1	2	3	4	5	6	7	8	9	10	11	12	13
5	FR5	DSM					X	C	C	N	N			N	N
6	FR6	FTA					N	X	N	N	N			C	C
7	FR7	RBD					S	C	X	N	N			S	C
8	FR8	FMEA					N	N	N	X	N			N	N
9	FR9	TRF					N	C	N	N	X			C	N
10	FR10	ETP													
11	FR11	OPT													
12	FR1	DRS					N	C	N	N	S			X	N
13	FR2	PLS					N	N	C	C	N			S	X

POST-DEPLOYMENT VALIDATION OF FRAMEWORK

GAURDS Validation Framework

The GUARDS validation strategy considers both short-term and long-term objectives. The main validation components in GAURDS depicted in the diagram are: (1) formal verification, (2) model-based evaluation, (3) fault injection, and (4) the methodology and the supporting toolset for schedulability analysis. Once the AIDF is validated with DFSS IDOV process, we can insert the AIDF architectural framework into the GAURDS validation framework for continuous validation, as more commercial off-the-shelf components are added and validated, as need, in the future (Fig. 4.31).

AIDF Generic Architecture for Real-time Upgradeable Dependable Systems
(GAURDS) Validation Framework
for the AIDF case study KBE SoS application

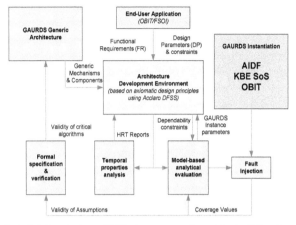

Instantiated mechanisms, COTS/unique components, modules for KBE system

Fig. 4.31.GAURDS validation framework for post-deployment validation.

SUMMARY

In the case study chapter, we have introduced the application, configuration, implementation, optimization, and verification, using the AIDF architecture framework for
optical backplane engineering. In the application section, we described the Free-Space
Optical Interconnect (FSOI) components and how to apply knowledge engineering allowing domain experts to capture their tacit and explicit knowledge on components and design rationale, using methods such as web-enabled DSM templates, rules, and ontologies.
In the configuration section, we showed various levels of granularity in analyzing the
AIDF, focusing at high-level down to engine module and element interaction analysis. In
the implementation section, we provided screenshots using Acclaro DFSS design matrix
architecture development tool. In the optimization and verification section, we show how

the DFSS IDOV, combined with the GAURDS validation framework, provides a pre-deployment and post-deployment validation strategy for the AIDF architectural framework

CHAPTER V: SYNERGISTIC VALIDATION METHODOLOGY

OVERVIEW

In this chapter, we provide a methodology for validation for the architecture and the complex systems of a KBE System-of-Systems (SoS) application. In this chapter, we introduce the standard approach to engineering validation, followed by introducing the Synergistic Validation Methodology (SVM) we used as our comprehensive approach to validation for the AIDF. This AIDF-SVM is shown to be divided into four divisions addressing multiple areas that correspond to well-defined methodologies in terms of standards, techniques, and methods employed today: *(I) Software Architecture, (II) Design Process, (III) Artificial Intelligence Inference Mechanisms,* and *(IV) Global Knowledge Acquisition Process.* The primary validation target is the software architecture, since the artificial intelligence design framework itself is an architectural framework. However, for comprehensiveness, the other three validation targets areas are also addressed in order to develop a synergistic methodology to configure a class of KBE systems having AI and Semantic Web elements based on the AIDF architectural framework platform. We provide a rational for each division and area of validation. Supportive validation research work, such as terminological surveys, best practices for decision support were conducted, as well as peer-reviewed development work with NASA Marshall Space Flight Center over two years. We concentrate our validation on the software architecture division, which forms the bedrock for development of the AIDF.

VALIDATION APPROACH FOR KBE SYSTEM-OF-SYSTEMS

STANDARD APPROACH TO ENGINEERING VALIDATION

Validation, Verification, and Requirement Specifications

The ultimate goal of *validation* is making sure that the "right" system is developed for the end-user, usually the customer, with respect to meeting *real-world needs*, whereas the goal of *verification* is making sure that the system was developed "right", with respect to meeting the engineering *requirement specifications* [Hoppe, 1993]. By assuring end-user formal and informal needs are clearly identified, hierarchical decomposition of system requirements significantly increase the likelihood of achieving design success during validation and verification (Table 5.1). When the engineering requirement specifications actually meet the end-user real-world needs, a *validated and verified* system can be deployed [Schulmeyer and MacKenzie , 1999]. In the case of software development, design and functional specifications must be met for verification [Jayaswal and Patton, 2006].

Table 5.1 Validation and verification of requirement specifications.

Validation Terminology	Description
Requirement Specifications	Assures end-user formal and informal needs are clearly identified
Validation	Assures the "right" system is built to meet real-world end-user needs
Verification	Assures the system is built "right" to meet engineering specifications

Validation Subsumes Verification of Requirement Specifications

Validation usually subsumes verification, so that system requirements are checked to meet engineering specifications *before* checking to make sure the system meets end-user needs. Many subsystems can also be independently verified and validated before

system integration and final validation (Fig. 5.1).

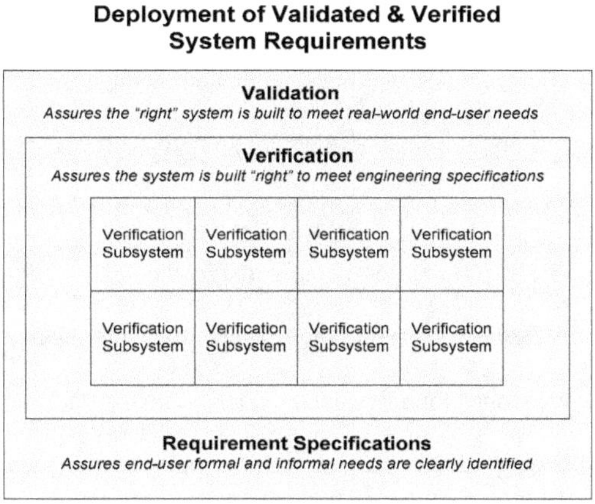

**Deployment of Validated & Verified
System Requirements**

Fig. 5.1. Validation and verification of system requirements.

Architectural Framework Improving on Time, Quality, Cost for KBE Development

In engineering, *timeliness*, *quality*, and *cost*, which have been widely accepted by

engineers to be central performance metrics [Muirhead and Simon, 1999]. An ideal ar-

chitectural framework should be able to improve on speed, cost, and time for KBE devel-

opment which allows for faster development time, higher quality with pre-deployment

and post-deployment validation strategy, and less expensive production for all KBE sys-

tems configured down the line. A validated, reconfigurable, and scalable architectural

framework having a high Return On Investment (ROI) for a KBE implementation is able

to deliver with respect these three attributes.

THESIS APPROACH TO COMPREHENSIVE VALIDATION BY STRATIFICATION

In this thesis, we develop a validated architectural framework that functions as a reconfigurable platform for design process automation with artificial intelligence leveraging intelligent agents on the Semantic Web. In order to achieve comprehensive validation, we recognized the need to research the state-of-the-art standards, techniques, and methods employable by KBE systems configured by the AIDF. For comprehensive validation, we *stratified* our approach into three layers: divisions, areas, and methodologies (Fig. 5.2). First, we scoped four divisional targets. Then, we decomposed the divisions into their corresponding validation areas, each having respective methodologies comprised of employable standards, techniques, and methods.

The *four divisional* targets comprised of multiple application areas correspond to *software architecture, design process, artificial intelligence, and knowledge acquisition*. The associated methodologies for each area were explored and distilled, with the resulting state-of-the-art standards, techniques, and methods identified and described in this chapter, reserving other aspects of validation to other chapters. This attempt at a seamless approach to validation, chapter by chapter, provides a structure to describe the topics with uniformity and consistency. The AIDF-SVM approach to validation is provided as a framework to provide methodologies in the future, as they appear and become accepted as standards. It is developed as a parallel approach that is recommended for application during the development process of a KBE SoS application based on the AIDF.

Validation of Identified Target Areas for Automated Product Design Support based on Artificial Intelligence and Web Services			
AIDF Validation Target I	AIDF Validation Target II	AIDF Validation Target III	AIDF Validation Target IV
Software Architecture	**Engineering Product Design Process**	**Artificial Intelligence Inference Mechanisms**	**Global Knowledge Acquisition Process**
Areas addressed	*Areas addressed*	*Areas addressed*	*Areas addressed*
(A) Architecture development process (B) Architecture framework model (C) Architectural structural/dynamic model (D) Architectural process model	(A) Engineering product design phases (B) Prevailing engineering design methods currently available	(A) Automated Knowledge Analysis & Design Synthesis Approach (B) Knowledge-based Ontology Implementation (C) Knowledge-based Ontology Evaluation (D) Knowledge-based Weighted Rules Verification (E) Knowledge-based Algorithmic Methods	(A) Authentication of Remote Knowledge Repositories (B) Intelligent Agents on Semantic Web
Set of Global Standards and Techniques Identified			

Fig. 5.2. Summary of identified validation targets and areas addressed.

VALIDATION THEME MAPPING CHAPTER BY CHAPTER

A common validation theme (Table 5.2) is threaded through each of the follow key parts of the thesis in a way that can be easily mapped: State-of-art-chapter, AIDF chapter, Case Study chapter, the Appendices, and the References. By knitting together key chapters by a stratified nomenclature (roman numeral, letter, and number), we provide a means a) to map the validation concepts *introduced* in the State-of-the-Art chapter; b) to the validation concepts *considered* during software architecture development; c) to the validation rationale *provided* in the Validation Chapter; d) to validation *demonstrated* by the configured architecture application in the Case Study Chapter; and e) to the validation content *expansion* provided in Appendix A. Hence, we emphasize the importance of validation as a basis for building a large-scale system based on an architectural frame-

work with the identified four target divisions of validation.

Table 5.2. Validation theme throughout chapters.

Chapter	Validation Theme
Chapter State of Art	Validation concepts *introduced*
Chapter AIDF	Validation concepts *considered* during software architecture development
Chapter Validation	Validation rationale *provided* in the Validation Chapter
Chapter Case Study	Validation *demonstrated* by the configured architecture application
Appendices	Validation methods *described* for architecture

VALIDATION GOALS FOR EACH CHAPTER

Validation Goal for State-of-the-art Chapter

The goal of State-of-the-art Chapter is to scope out the validation approach into

four target divisions, each having associated application areas that have validation meth-

odologies in terms of state-of-the-art standards, techniques, and methods employable to-

day. Of the four validation targets, we introduce the *primary* validation target as the

software architecture, since the AIDF itself is an architectural framework.

Validation Goal for Artificial Intelligence Chapter

The goal of the AIDF chapter is to introduce and describe the AIDF architectural

framework by hierarchical decomposition of requirements specifications and correspond-

ing design parameters as a basis for verification of meeting engineering requirements.

The AIDF architectural framework model and the design process model are introduced,

along with the twenty modules in the AI engine block providing automated design support. Each of the modules in the dual engine housing the design modules and AI modules is knowledge engineered, so that domain-specific knowledge is captured in a format that a machine can process for computability. The implementation of the case study architectural framework is reserved for the case study chapter, along with a demonstration of the validation application.

Validation Goal for Case Study Chapter Objective

The Case Study Chapter provides a means to *demonstrate* how the validation methodologies were applied. The actual implementation of the validation methodology will be demonstrated in the Case Study Chapter. The structure of this chapter is designed to allow the reader to easily map the introduced topics to the Validation Chapter, which describes show how each of the validation target areas and associated methodologies introduced in this chapter, in terms of standards, techniques, and domain-specific methods, are meshed together into a unified whole applied to the case study.

Validation Goal for Synergistic Validation Methodology Chapter

Our comprehensive approach to validation meshing all of these topics together into a cohesive whole is reserved for the Validation Chapter, where we introduce our cohesive approach to validation, weaving together the standards, techniques, and methods applied to each application area corresponding to the validation target divisions. We refer to our validation approach for the AIDF as the Synergistic Validation Methodology (SVM). The AIDF-SVM describes validation of the implementation, the rationale for the synergistic validation approach for the AIDF platform per area, and the survey and test prototyping done as part of the preliminary research. While the other validation divisions are important for a functioning KBE SoS, validation of the software architecture is the main concentration in this chapter. Thus, we introduce the SVM as a recommended approach for engineers to validate a class of scalable KBE SoS applications for automated design support, which can be configured based on our AIDF platform. As an implementation, Acclaro Design for Six Sigma, an industry-grade architecture development tool acquired by NASA funds, is used for both configuration and validation.

Validation Goal of Appendix

Since the primary validation target for the AIDF is the software architecture, we describe the applied validation rationale for the software architecture in Appendix A, in terms of specific methods and decisions considered during AIDF model development.

VALIDATION CHAPTER APPROACH

Comprehensive Validation by Target Division

We have divided the comprehensive approach (Fig. 5.2) to validation into four validation targets treated separately in this chapter, and tied together into a unified whole in the validation chapter. This forms a basis to describe the applied validation techniques on the case study implementation, detailed in the Case Study Chapter. Each validation target can be decomposed into areas, each having respective methodologies comprised of employable standards, techniques, and methods. In this chapter, we will identify and describe each individual topic separately, as an introduction the concepts expanded on and applied later in the validation chapter and case study chapter.

Primary Validation Target is the Software Architecture

Each validation target can be decomposed into areas, each having respective methodologies comprised of employable standards, techniques, and methods. Hence, for the sake of comprehensive validation, we divided the approach to validation by scoping four fundamental targets: *(I) Software Architecture, (II) Design Process, (III) Artificial Intelligence Inference Mechanisms,* and *(IV) Global Knowledge Acquisition Process.* Although much work on validation is done addressing as many areas as possible associated with the AIDF, the primary validation target is the software architecture, since the artificial intelligence design framework itself is an architectural framework. Thus, validation of the implementation concentrates primarily on the framework. However, for comprehensiveness, the other three validation targets areas are also addressed in order to develop a synergistic methodology to configure a class of KBE systems having AI and Semantic Web elements based on the AIDF architectural framework platform.

Structure of the Validation Chapter

We provide a section as an overview of the validation target areas, followed by two sections identifying and describing the employable state-of-the-art standards, techniques, and methods, respectively. Each of these validation areas can be mapped to the validation chapter, where we show how each of these state-of-the-art standards, techniques, and methods can be applied in the validation of the thesis implementation. The rationale and application of the validation standards, techniques, and methods discussed in this chapter are described in more detail in the Validation Chapter and the Case Study Chapter, respectively.

**Overview of Comprehensive Validation Approach for
AIDF KBE System Launching Platform**

*Comprehensive
Validation Approach*

**Artificial Intelligence Design Framework
Synergistic Validation Methodology (AIDF-SVM)**

Vertical label (left): **Architectural Framework Axiomatic Implementation & Six Sigma Validation**

Acclaro DFSS Implementation of AIDF architectural framework model with Front-End Validation Techniques
*Using Acclaro Design for Six Sigma (DFSS), industry-grade architectural development tool
acquired by NASA funds and used by General Dynamics*

Acclaro DFSS *AIDF Platform Validation* Axiomatic Design Theory Complexity Reduction	**Acclaro DFSS** *AIDF Platform Validation* Design Structure Matrix Dependency Resolution	**Acclaro DFSS** *AIDF Platform Validation* Failure-Mode & Effects Analysis	**Acclaro DFSS** *AIDF Platform Validation* Quality Function Deployment

Validation Methodology for AIDF Launching Platform using AIDF-SVM
*Development of the AIDF Synergistic Validation Methodology (SVM)
for the Reconfigurable AIDF Platform Launching Scalable KBE Systems*

Vertical label (left): **Synergistic Validation Methodology (AIDF-SVM) for Launching Scalable KBE Systems with Reconfigurable AIDF Platform**

AIDF-SVM Validation Target Areas
addressed by implementing a set of associated standards and techniques

AIDF Validation Target I **Software Architecture**	*AIDF Validation Target II* **Engineering Product Design Process**	*AIDF Validation Target III* **Artificial Intelligence Inference Mechanisms**	*AIDF Validation Target IV* **Global Knowledge Acquisition Process**
Areas addressed (A) Architecture development process (B) Architecture framework model (C) Architectural structural/dynamic model (D) Architectural process model	*Areas addressed* (A) Engineering product design phases (B) Prevailing engineering design methods currently available	*Areas addressed* (A) Automated Knowledge Analysis & Design Synthesis Approach (B) Knowledge-based Ontology Implementation (C) Knowledge-based Ontology Evaluation (D) Knowledge-based Weighted Rules Verification (E) Knowledge-based Algorithmic Methods	*Areas addressed* (A) Authentication of Remote Knowledge Repositories (B) Intelligent Agents on Semantic Web

**Set of Global Standards and Techniques
Implemented, employed, and adapted for AIDF-SVM**

Preliminary Validation Research

Vertical label (left): **Research Survey & Prototyping**

Best Practices Survey of Decision Support Systems	Best Practices Survey on Knowledge-Based System Validation Methods	Best Practices Survey of Automated Design Process Support Systems	Test Prototyping of Knowledge base for Intelligent Agent application on Semantic Web

Fig. 5.3. Comprehensive Validation Approach for KBE SoS based on AIDF.

IDENTIFICATION OF VALIDATION TARGET DIVISION BY APPLICATION AREA

Identification of Main Validation Target Divisions

For the sake of comprehensive validation, we divided the comprehensive approach to validation by scoping four target divisions (Table 5.3): *(I) Software Architecture, (II) Design Process, (III) Artificial Intelligence Inference,* and *(IV) Semantic Web-enabled Knowledge Acquisition Process.*

Table 5.3. Validation target divisions.

Validation Target Division
(I) SOFTWARE ARCHITECTURE
(II) PRODUCT DESIGN PROCESS
(III) ARTIFICIAL INTELLIGENCE
(IV) KNOWLEDGE ACQUISITION

Software Architecture Division Validation Areas

Validation of the *Software Architecture division*, the primary validation target division (Table 5.4), addresses four application areas: (A) *architecture development process*, (B) *architecture framework model*, (C) *architectural structural/dynamic model*, (D) and *architectural process model*.

Table 5.4. Validation areas for target division I: Software Architecture.

Validation Target Area	Appendix
(A) Architecture development process	I-A
(B) Architecture framework model	I-B
(C) Architectural structural/dynamic model	I-C
(D) Architectural process model	I-D

Design Process Division Validation Areas

Validation of the *Design Process division* addresses two application areas (Table 5.5): (A) systems engineering *product design phases,* and (B) prevailing *engineering design methods* currently available.

Table 5.5 Validation areas for target division II: Design Process.

Validation Target Area
(A) Systems engineering *product design phases*
(B) Prevailing *engineering design methods*

Artificial Intelligence Division Validation Areas

Validation of the *Artificial Intelligence division* addresses five application areas (Table 5.6): *(A) Automated Knowledge Analysis and Design Synthesis Approach, (B) Knowledge-based Ontology Implementation, (C) Knowledge-based Ontology Evaluation, (D) Knowledge-based Weighted Rules Verification,* and *(E) Knowledge-based Algorithmic Methods.*

Table 5.6 Validation areas for target division III: Artificial intelligence.

Validation Target Area
(A) Automated Knowledge Analysis and Design Synthesis Approach
(B) Knowledge-based Ontology Implementation
(C) Knowledge-based Ontology Evaluation
(D) Knowledge-based Weighted Rules Verification
(E) Knowledge-based Algorithmic Methods

Knowledge Acquisition Validation Areas

Validation of the *Knowledge Acquisition division* addresses two application areas

(Table 5.7): *(A) Authentication of Remote Knowledge Repositories, (B) Intelligent Agents*

on Semantic Web.

Table 5.7. Validation areas for target division III: Knowledge acquisition.

Validation Target Area
(A) Authentication of Remote Knowledge Repositories
(B) Intelligent Agents on Semantic Web

VALIDATION RESEARCH STRATIFICATION PROCESS

For comprehensive validation, we stratified our approach into three layers: divi-

sions, areas, and methodologies. In summary, we showed the table for each target divi-

sion and their corresponding application area. In the next section we will further elabo-

rate on this table by identifying the corresponding methodologies to the application areas,

in terms of employable standards, techniques, and methods.

Stratification by Target Division and Area

First, we scoped four divisional targets. Then, we decomposed the divisions into

their corresponding validation areas, each having respective methodologies comprised of

employable standards, techniques, and methods. The validation divisions are: *(I) Soft-*

ware Architecture, (II) Design Process, (III) Artificial Intelligence Inference Mecha-

nisms, and *(IV) Global Knowledge Acquisition Process.* Validation of the *Software Ar-*

chitecture division, the primary validation target division, addresses four application areas: (A) *architecture development process*, (B) *architecture framework model*, (C) *architectural structural/dynamic model*, (D) and *architectural process model*. Validation of the *Design Process division* addresses two application areas: (A) systems engineering *product design phases,* and (B) prevailing *engineering design methods* currently available. Validation of the *Artificial Intelligence division* addresses five application areas: *(A) Automated Knowledge Analysis and Design Synthesis Approach, (B) Knowledge-based Ontology Implementation, (C) Knowledge-based Ontology Evaluation, (D) Knowledge-based Weighted Rules Verification,* and *(E) Knowledge-based Algorithmic Methods.* Validation of the Semantic-Web enabled *Knowledge Acquisition division* addresses two application areas: *(A) Authentication of Remote Knowledge Repositories, (B) Intelligent Agents on Semantic Web.*

Stratification by Target Area and Methodology

As mentioned, each validation *target* can be decomposed into application *areas*, each having respective validation *methodologies* comprised of employable standards, techniques, and methods. In this chapter, we will briefly introduce each term by *identifying* and *describing* the associated methodologies in terms of standards, methods, and techniques employed in the validation of the AIDF architectural framework and its associated KBE system and Semantic Web elements. In this section, we will identify each application area for validation and associated methodology in terms of standards, techniques, and methods by listing them with corresponding citations. In the next section, we will provide a description of each of these standards, methods, and techniques identified and listed. These descriptions are only intended to provide a brief explanation capturing the essence of the topics addressed, as background material. The diligent reader is directed to the thesis Appendices and references, or for further expansion, to the current literature available online for a more detailed treatment of the subject matter introduced in this state-of-the-art chapter.

VALIDATION DIVISION STRATIFICATION BY AREA AND METHODOLOGY

Division Target I: Software Architecture Area Validation Methodology

In order to validate the primary validation target of the AIDF involving *Software Architecture*, four areas were identified (Table 5.8) that addressed the (A) *architecture development process*, (B) *architecture framework model*, (C) *architectural structural/dynamic model*, (D) and *architectural process model*.

(A) Validation of Architectural Development Process:

For validation of the area addressing the *architectural development process*, we identified techniques and standards recommended by *(1) ANSI/IEEE Std. 1471 terminology, (2) National Defense Authorisation Act for Fiscal Year 1996 (Clinger-Cohen Act), (3) OMG-MDA technology independence approach, (4) First International Workshop on IT Architectures for Software Systems in 1995, (5) Shaw-Carnegie Mellon architectural categorizations defining framework model, structural model, dynamic model, process model, functional model, (6) ATAM for architectural trade-off analysis, (7) CBAM for architectural cost assessment, (8) Acclaro Design for Six Sigma (DFSS) implementation, (9) Telelogic TAU implementation of UML and SysML, (10) Rational Unified Process (RUP) (11) Architecture-driven software construction (12) Waterfall Model (13) Spiral*

Model (14) Capability Maturity Model Integration (CMMI) (15) Rapid Application De-

velopment (RAD) (16) Dynamic Systems Development Method

Table 5.8. Validation target area I-A: Software development process.

Methodology in terms of Standards, Techniques, Methods	Appendix	Key References	Date
(1) ANSI/IEEE Std 1471 terminology	I-A-1	IEEE ANSI/IEEE Std 1471	2006
(2) National Defense Authorisation Act for Fiscal Year 1996 (Clinger-Cohen Act)	I-A-2	U.S. Govt. Clinger-Cohen Act	1996
(3) OMG-Model-Driven Architecture (MDA) technology independence approach	I-A-3	Object Management Group	2006
(4) First International Workshop on IT Architectures for Software Systems in 1995	I-A-4	Garlan	1995
(5) Shaw-Carnegie Mellon architectural categorizations	I-A-5	Shaw	1996
(6) ATAM for architectural trade-off analysis	I-A-6	Bass	2003
(7) CBAM for architectural cost assessment	I-A-7	Bass	2003
(8) Acclaro Design for Six Sigma (DFSS) implementation	I-A-8	Axiomatic Design, Inc.	2006
(9) Telelogic TAU implementation of UML and SysML	I-A-9	Telelogic, Inc.	2006
(10) Rational Unified Process (RUP)	I-A-10	Rational Software Corporation, IBM, Inc.	2006
(11) Architecture-driven software construction	I-A-11	Sewell	2002
(12) Waterfall Model	I-A-12	DOD-STD-2167A/498 Handbook	1995
(13) Spiral Model	I-A-13	Boehm	1988
(14) Capability Maturity Model Integration (CMMI)	I-A-14	Software Engineering Institute (SEI)	2006
(15) Rapid Application Development (RAD)	I-A-15	IBM, Inc.	2006
(16) Dynamic Systems Development Method	I-A-16	Agile Alliance	2006

(B) Validation of Architecture Framework Model

For validation of the area (Table 5.9) addressing the *architecture framework model*, we identified techniques and standards recommended by *(1) AIDF implementation/Acclaro Design for Six Sigma, (2) Failure Mode and Effects Analysis (FMEA), (3) Design Structure Matrix (DSM) dependency resolution, (4) Axiomatic Design Theory (ADT) risk mitigation, (5) Quality Function Deployment (QFD) (6) Generic Architecture for Upgradeable Real-Time Dependable Systems (GAURDS) validation framework (6) Best Practices based on survey of 25 decision support tools and 10 design frameworks, (7) National Academy for Engineering state-of-the-art 2001 report outlining Approaches to Engineering Design*

Table 5.9.Validation target area I-B: Architecture framework model.

Methodology in terms of Standards, Techniques, Methods	Appendix	Key References	Date
(1) AIDF implementation/Acclaro Design for Six Sigma (DFSS)	I-B-1	Axiomatic Design, Inc.	2006
(2) Failure Mode and Effects Analysis (FMEA)	I-B-2	Six Sigma, Motorola	2006
(3) Design Structure Matrix (DSM) dependency resolution	I-B-3	Albin	2003
(4) Axiomatic Design Theory (ADT) risk mitigation	I-B-4	Suh	2001
(5) Quality Function Deployment (QFD)	I-B-5	Six Sigma Motorola	2006
(6) Generic Architecture for Upgradeable Real-Time Dependable Systems (GAURDS) validation framework	I-B-6	European Strategic Program on Research in Information Technology (ESPRIT)	1999
(7) National Academy for Engineering state-of-the-art 2001 report outlining Approaches to Engineering Design	I-B-7	National Research Council	2001

(C) Validation of Architectural Structural/Dynamic Model

For validation of the area (Table 5.10) addressing the *structural/dynamic models*, we identified techniques and standards recommended by *(1) High Level Integrated Design Environment (HIDE) for dependability (2) Telelogic TAU SysML implementation/code validation/verification.*

Table 5.10. Validation target area I-C: Architectural structural/dynamic model.

Standards, Techniques, Methods	Detail	Key Reference	Date
(1) High Level Integrated Design Environment (HIDE) for dependability	I-C-1	Bondavalli	2003
(2) Telelogic TAU UML/SysML implementation	I-C-2	Telelogic, Inc.	2006
(3) NASA SATC metrics for object-oriented code evaluation in terms of classes, methods, cohesion, and coupling	I-C-3	Software Assurance Technology Center (SATC) at NASA Goddard Space Flight Center	1998

(D) Validation of Process Model

For validation of the area addressing (Table 5.11) the (D) *process model*, we identified techniques and standards recommended by *(1) Axiomatic V-Model mapping process (2) National Academy of Engineering.*

Table 5.11. Validation target area I-D: Architectural structural/dynamic model.

Standards, Techniques, Methods	Detail	Reference	Date
(1) Axiomatic V-Model mapping process	I-D-1	Suh	2001
(2) National Academy of Engineering	I-D-2	National Research Council	2001

Summary of Software Architecture Validation Division

For the Software Architecture validation target (Table 5.12), we summarize the

corresponding validation application areas and their associated validation methodology

standards, techniques, and methods available. In order to validate the primary validation

target of the AIDF involving *Software Architecture,* four areas were identified that ad-

dressed the (A) *architecture development process*, (B) *architecture framework model*, (C)

architectural structural/dynamic model, (D) and *architectural process model.*

For validation of the area addressing the *architectural development process*, we

identified techniques and standards recommended by *(1) ANSI/IEEE Std. 1471 terminol-

ogy, (2) National Defense Authorisation Act for Fiscal Year 1996 (Clinger-Cohen Act),

(3) OMG-MDA technology independence approach, (4) First International Workshop on

IT Architectures for Software Systems in 1995, (5) Shaw-Carnegie Mellon architectural

categorizations defining framework model, structural model, dynamic model, process

model, functional model, (6) ATAM for architectural trade-off analysis, (7) CBAM for

architectural cost assessment, (8) Acclaro Design for Six Sigma (DFSS) implementation,

(9) Telelogic TAU implementation of UML and SysML, (10) Rational Unified Process

(RUP) (11) Architecture-driven software construction (12) Waterfall Model (13) Spiral

Model (14) Capability Maturity Model Integration (CMMI) (15) Rapid Application De-

velopment (RAD) (16) Dynamic Systems Development Method*

For validation of the area addressing the *architecture framework model,* we iden-

tified techniques and standards recommended by *(1) AIDF implementation/Acclaro De-

sign for Six Sigma, (2) Failure Mode and Effects Analysis (FMEA), (3) Design Structure

Matrix (DSM) dependency resolution, (4) Axiomatic Design Theory (ADT) risk mitiga-*

tion, (5) Quality Function Deployment (QFD) (6) Generic Architecture for Upgradeable

Real-Time Dependable Systems (GAURDS) validation framework (6) Best Practices

based on survey of 25 decision support tools and 10 design frameworks, (7) National

Academy for Engineering state-of-the-art 2001 report outlining Approaches to Engineer-

ing Design.

For validation of the area addressing the *structural/dynamic models*, we identified

techniques and standards recommended by *(1) High Level Integrated Design Environ-*

ment (HIDE) for dependability (2) Telelogic TAU SysML implementation/code valida-

tion/verification.

For validation of the area addressing the (D) *process model*, we identified tech-

niques and standards recommended by *(1) Axiomatic V-Model mapping process (2) Na-*

tional Academy of Engineering.

Table 5.12. Software architecture validation: Standards, techniques, methods.

Division	Application Areas	Validation Methodology standards, techniques, and methods
(I) Software Architecture	(A) Architecture development process	*(1) ANSI/IEEE Std 1471 (2) Clinger-Cohen Act (3) OMG-MDA (4) First International Workshop on IT Architectures for Software Systems in 1995, (5) Shaw-Carnegie Mellon architectural categorizations (6) ATAM for architectural trade-off analysis, (7) CBAM for architectural cost assessment, (8) Acclaro Design for Six Sigma (DFSS) implementation, (9) Telelogic TAU implementation of UML and SysML, (10) Rational Unified Process (11) Architecture-Driven Software Construction*
	(B) Architecture framework model	*(1) AIDF implementation/Acclaro Design for Six Sigma, (2) Failure Mode and Effects Analysis (FMEA), (3)Design Structure Matrix (DSM) dependency resolution, (4) Axiomatic Design Theory (ADT) risk mitigation, (5) Quality Function Deployment (QFD) (6) Generic Architecture for Upgradeable Real-Time Dependable Systems (GAURDS) validation framework (7) Survey of 25 decision support tools and 10 frameworks*
	(C) Architectural structural/dynamic model	*(1) High Level Integrated Design Environment (HIDE) for dependability (2) Telelogic TAU SysML implementation/code validation/verification*
	(D) Architectural process model	*(1) Axiomatic V-Model mapping process (2) National Academy of Engineering 2001 Report on Approaches to Engineering Design*

Division Target II: Design Process Area Validation Methodology

In order to validate the second validation target of the AIDF involving *Design Process*, two areas were identified that addressed the (A) systems engineering *product design phases* and (B) prevailing *engineering design methods* currently available.

(A) Validation of Product Design Phases

For validation of the area (Table 5.13) addressing the systems engineering *product design phases,* we identified the widely-accepted characterization for engineering design phases as defined by *Pahl and Beitz.*

Table 5.13. Validation target area II-A: Product design phases.

Standards, Techniques, Methods	Key References	Date
(1) Pahl and Beitz	Pahl and Beitz	1988

(B) Validation of Engineering Design Methods

For validation of the area addressing the prevailing engineering *design methods* currently available, we identified standards and techniques recommended by *(1) Axiomatic Design Theory (ADT), (2) Theory of Inventive Problem Solving (TRIZ), (3) Hierarchical Multi-layer Design (MLH), (4) Quality Function Deployment (5) Design Structure Matrix (DSM) (6) Fault-Tree-Analysis (FTA) (7) Failure-Mode and Effects Analysis (8) Reliability Block Diagram (RBD) Analysis (9)Technology Risk Factor (TRF) Assessment (10) Entropy (ETP) Analysis, and (11) Case Study: Optical Backplane Engineering (OPT) Domain.*

Table 5.14. Validation target area II-B: Design methods.

Standards, Techniques, Methods	Key References	Date
(1) Axiomatic Design Theory (ADT)	Suh	2001
(2) Theory of Inventive Problem Solving (TRIZ)	Chai	2005
(3) Hierarchical Multi-layer Design (MLH)	Trewn	2000
(4) Quality Function Deployment	Six Sigma, Motorola	2006
(5) Design Structure Matrix (DSM)	Six Sigma, Motorola	2006
(6) Fault-Tree-Analysis (FTA)	Systems and Reliability Research, Nuclear Regulatory Commission (NUREG -0492)	1984
(7) Failure-Mode and Effects Analysis	Six Sigma, Motorola	2006
(8) Reliability Block Diagram (RBD) Analysis	NASA Johnson Space Center (JSC)	1995
(9)Technology Risk Factor (TRF) Assessment	MIT/NASA Masters Thesis/Brady	2002
(10) Entropy (ETP) Analysis	Shannon	1948
(11) Case Study: Optical Backplane Engineering (OPT) Domain	Grimes	2005

Summary of Design Process Validation Division

For the Design Process validation target, we summarize the corresponding validation application areas and their associated validation methodology standards, techniques, and methods available (Table 5.15). In order to validate the second validation target of the AIDF involving *Design Process*, two areas were identified that addressed the (A) systems engineering *product design phases* and (B) prevailing *engineering design methods* currently available. For validation of the area addressing the systems engineering *product design phases*, we identified the widely-accepted characterization for engineering design phases as defined by *Pahl and Beitz*. For validation of the area addressing the prevailing engineering *design methods* currently available, we identified standards and techniques recommended by *(1) Axiomatic Design Theory (ADT), (2) Theory of Inventive Problem Solving (TRIZ), (3) Hierarchical Multi-layer Design (MLH), (4) Quality Function Deployment (5) Design Structure Matrix (DSM) (6) Fault-Tree-Analysis (FTA) (7) Failure-Mode and Effects Analysis (8) Reliability Block Diagram (RBD) Analysis (9)Technology Risk Factor (TRF) Assessment (10) Entropy (ETP) Analysis, and (11) Case Study: Optical Backplane Engineering (OPT) Domain.*

Table 5.15. Design process validation: Standards, techniques, methods.

Division	Application Areas	Validation Methodology standards, techniques, and methods
(II) Design Process	(A) Engineering product design phase	*Methodology by Pahl and Beitz*
	(B) Prevailing engineering design methods currently available	*(1) Axiomatic Design Theory (ADT), (2) Theory of Inventive Problem Solving (TRIZ), (3) Hierarchical Multi-layer Design (MLH), (4) Quality Function Deployment (5) Design Structure Matrix (DSM) (6) Fault-Tree-Analysis (FTA) (7) Failure-Mode and Effects Analysis (8) Reliability Block Diagram (RBD) Analysis (9)Technology Risk Factor (TRF) Assessment (10) Entropy (ETP) Analysis, and (11) Case Study: Optical Backplane Engineering (OPT) Domain*

Division Target III: Artificial Intelligence Area Validation Methodology

In order to validate the third validation target of the ADIF involving *Artificial Intelligence*, five areas were identified that addressed the *(A) Automated Knowledge Analysis and Design Synthesis Approach, (B) Knowledge-based Ontology Implementation, (C) Knowledge-based Ontology Evaluation, (D) Knowledge-based Weighted Rules Verification*, and *(E) Knowledge-based Algorithmic Methods.*

(A) Validation of Automated Knowledge Analysis and Design Synthesis Approach

For validation of the *Automated Knowledge Analysis and Design Synthesis* approach to knowledge-based engineering approach of the AIDF, we identified the standards and techniques (Table 5.16) recommended by *(1) the de-facto standard for knowledge modeling as defined by Common Knowledge Analysis and Design Synthesis (CommonKADS) Approach, and the (2) the European Union Esprit II VALID project exploring the state-of-the-art for validation of KB systems.*

Table 5.16. Validation target area III-A: Knowledge analysis and design synthesis.

Standards, Techniques, Methods	Key References	Date
(1) the de-facto standard for knowledge modeling as defined by Common Knowledge Analysis and Design Synthesis (CommonKADS) Approach	European Strategic Program on Research in Information Technology (ESPRIT)	2006
(2) the European Union Esprit II VALID project exploring the state-of-the-art for validation of KB systems	ESPRIT II VALID Project	1993

(B) Validation of Knowledge-based Ontology Implementation

For validation of the *Knowledge Base Ontology Implementation*, we identified the standards and techniques (Table 5.17) provided by *(1) Concept Map tools (CmapTools) for domain expert knowledge capture with ontologies, (2) Concept Map tools (CmapTools) Methodology to connect to global knowledge repositories on Semantic Web by demonstrating the exportability of ontology constructs into XML and Web Ontology Language (OWL) for use by intelligent agents.*

Table 5.17. Validation target area III-B: KB ontology Implementation.

Standards, Techniques, Methods	Key References	Date
(1) Concept Map tools (CmapTools) to test for domain expert knowledge capture with ontologies	Florida Institute for Human and Machine Cognition (IHMC)	2006
(2) Concept Map tools (CmapTools) Methodology to demonstrate how to test for connectivity to global knowledge repositories on Semantic Web by demonstrating the exportability of ontology constructs into XML and Web Ontology Language (OWL) for use by intelligent agents	Florida Institute for Human and Machine Cognition (IHMC)	2006

(C) Validation of Knowledge-based Ontology Evaluation

For validation of the *Knowledge Base Ontology Evaluation* metrics, we identified the

standards and techniques (Table 5.18) provided by the state-of-art research on ontology

validation based on mathematical assessment of ontology quality and quantity [Ontology,

2006] in terms of *breadth, fan-out,* and *tangledness.*

Table 5.18. Validation target area III-C: KB Ontology evaluation.

Standards, Techniques, Methods	Key References	Date
(1) mathematical assessment of ontology quality and quantity in terms of breadth, fan-out, and tangledness	Ontology evaluation and validation (Laboratory for Applied Ontology)	2006

(D) Validation of Knowledge-based Weighted Rules Verification

For validation of the *Knowledge-based Weighted Rules Verification,* we identified

the standards and techniques (Table 5.19) provided by *(1) verification with HP Labs Jena coding for inference engine, (2) code evaluation of Java Expert System Shell (JESS), (3) weighted rules verification with Matlab Fuzzy Logic Toolbox*

Table 5.19 Validation target area III-D: KB weighted rules verification

Standards, Techniques, Methods	Key References	Date
(1) verification with HP Labs Jena coding for inference engine	Hewlett-Packard Labs	2006
(2) code evaluation of Java Expert System Shell (JESS)	Sandia National Laboratories	2006
(3) weighted rules verification with Matlab Fuzzy Logic Toolbox	Matlab	2006

(E) Validation of Knowledge-based Algorithmic Methods.

For validation of the *Knowledge-based Algorithmic Methods,* we identified the standards and techniques (Table 5.20) provided by mechanisms for *(1) domain rule support (2) predicate logic support (3) algorithmic reasoning support (4) fuzzy logic support (5) neural network support (6) genetic algorithm support (7) Conant transmission support (8) Calibrated Bayesian Support (9) Data Mining Support.*

Table 5.20. Validation target area III-E: KB algorithmic methods.

Standards, Techniques, Methods	Key References	Date
(1) domain rule support	KBE/ICAD	2006
(2) predicate logic support	Norvig	2003
(3) algorithmic reasoning support	Norvig	2003
(4) fuzzy logic support	Mathworks Fuzzy Logic Toolbox	2006
(5) neural network support	Mathworks Neural Network Toolbox	2006
(6) genetic algorithm support	Mathworks Genetic Algorithms Toolbox	2006
(7) Conant transmission support	Conant	1976
(8) Calibrated Bayesian Support	Norsys Software Corp.	2006
(9) Data Mining Support	Pang	2005

Summary of Artificial Intelligence Division

For the Artificial Intelligence validation target division, we summarize the corresponding validation application areas and their associated validation methodology standards, techniques, and methods available (Table 5.21). In order to validate the third validation target of the ADIF involving *Artificial Intelligence*, five areas were identified that addressed the *(A) Automated Knowledge Analysis and Design Synthesis Approach, (B) Knowledge-based Ontology Implementation, (C) Knowledge-based Ontology Evaluation, (D) Knowledge-based Weighted Rules Verification,* and *(E) Knowledge-based Algorithmic Methods.*

For validation of the *Automated Knowledge Analysis and Design Synthesis* approach to Knowledge-Based Engineering approach of the AIDF, we identified the standards and techniques recommended by *(1) the de-facto standard for knowledge modeling as defined by Common Knowledge Analysis and Design Synthesis (CommonKADS) Approach, and the (2) the European Union Esprit II VALID project exploring the state-of-the-art for validation of KB systems.* For validation of the *Knowledge Base Ontology Implementation*, we identified the standards and techniques provided by *(1) Concept Map tools (CmapTools) for domain expert knowledge capture with ontologies, (2) Concept Map tools (CmapTools) Methodology to connect to global knowledge repositories on Semantic Web by demonstrating the exportability of ontology constructs into XML and Web Ontology Language (OWL) for use by intelligent agents.* For validation of the *Knowledge Base Ontology Evaluation* metrics, we identified the standards and techniques provided by the state-of-art research on ontology validation based on mathematical assessment of ontology quality and quantity in terms of *breadth, fan-out,* and *tangledness.*

For validation of the *Knowledge-based Weighted Rules Verification,* we identified the standards and techniques provided by *(1) verification with HP Labs Jena coding for inference engine, (2) code evaluation of Java Expert System Shell (JESS), (3) weighted rules verification with Matlab Fuzzy Logic Toolbox.* For validation of the *Knowledge-based Algorithmic Methods,* we identified the standards and techniques provided by mechanisms for *(1) domain rule support (2) predicate logic support (3) algorithmic reasoning support (4) fuzzy logic support (5) neural network support (6) genetic algorithm support (7) Conant transmission support (8) Calibrated Bayesian Support (9) Data Mining Support.*

Table 5.21. Artificial intelligence validation: Standards, techniques, methods.

Division	Application Areas	Validation Methodology standards, techniques, and methods
(III) Artificial Intelligence inference Mechanisms	(A) Automated Knowledge Analysis and Design Synthesis Approach	*(1) the de-facto standard for knowledge modeling as defined by Common Knowledge Analysis and Design Synthesis (Common-KADS) Approach, and the (2) the European Union Esprit II VALID project exploring the state-of-the-art for validation of KB systems*
	(B) Knowledge-based Ontology Implementation	*(1) Concept Map tools (CmapTools) to test for domain expert knowledge capture with ontologies, (2) Concept Map tools (Cmap-Tools) Methodology to test for connectivity to global knowledge repositories on Semantic Web by demonstrating the exportability of ontology constructs into XML and Web Ontology Language (OWL) for use by intelligent agents*
	(C) Knowledge-based Ontology Evaluation	Mathematical assessment of ontology quality and quantity in terms of *breadth, fan-out,* and *tangledness*
	(D) Knowledge-based Weighted Rules Verification	*(1) verification with HP Labs Jena coding for inference engine, (2) code evaluation of Java Expert System Shell (JESS), (3) weighted rules verification with Matlab Fuzzy Logic Toolbox*
	(E) Knowledge-based Algorithmic Methods	*(1) domain rule support (2) predicate logic support (3) algorithmic reasoning support (4) fuzzy logic support (5) neural network support (6) genetic algorithm support (7) Conant transmission support (8) Calibrated Bayesian Support (9) Data Mining Support*

Division Target IV: Knowledge Acquisition Area Methodology Validation

In order to validate the fourth validation target of the AIDF involving of valida-

tion of the *Global Knowledge Acquisition Process*, two areas were identified that ad-

dressed *(A) Authentication of Remote Knowledge Repositories* and *(B) Intelligent Agents*

on Semantic Web.

(A) Authentication of Remote Knowledge Repositories

For validation of the approach to *Authentication of Remote Knowledge Reposito-*

ries, we identified (Table 5.22) *Cluster-on-demand-architecture (Coda)* as the choice technol-

ogy.

Table 5.22 Validation target area IV-A: KB ontology evaluation

Standards, Techniques, Methods	Key Reference	Date
(1) Cluster-on-demand-architecture (Coda)	Satyanarayanan	1990

(B) Validation of Intelligent Agents on Semantic Web

For validation of the *intelligent agents on the Semantic Web*, we identified (Table

5.23) the application of the frontier technology of *DAMLJESSKB* and its evaluation and

verification methods.

Table 5.23 Validation target area IV-B: KB intelligent agents on semantic web

Standards, Techniques, Methods	Key Reference	Date
DAMLJESSKB evaluation and verification methods with Web Services	Kopena	2003

Summary of Knowledge Acquisition

For the Knowledge Acquisition validation target division, we summarize the corresponding validation application areas and their associated validation methodology standards, techniques, and methods available (Table 5.24). In order to validate the fourth validation target of the AIDF involving of validation of the *Global Knowledge Acquisition Process*, two areas were identified that addressed *(A) Authentication of Remote Knowledge Repositories* and *(B) Intelligent Agents on Semantic Web.*

For validation of the approach to *Authentication of Remote Knowledge Repositories,* we identified *Cluster-on-demand-architecture (Coda)* as the choice technology. For validation of the *intelligent agents on the Semantic Web*, we identified the application of the frontier technology of *DAMLJESSKB* and its evaluation and verification methods.

Table 5.24 Knowledge acquisition: Standards, techniques, methods

Division	Application Areas	Validation Methodology standards, techniques, and methods
(IV) Global Knowledge Acquisition Process	(A) Authentication of Remote Knowledge Repositories	*Cluster-on-demand architecture (Coda) for distributed computing with Web Services*
	(B) Intelligent Agents on Semantic Web	*Semantic Web: DAMLJESSKB* evaluation and verification methods

SUMMARY OF VALIDATION DIVISION TARGETING

In this thesis, we develop a validated platform for design process automation based on Semantic Web-enabled artificial intelligence. In order to achieve comprehensive validation, we recognized the need to research the state-of-the-art standards, techniques, and methods employable by KBE systems configured by the AIDF. For comprehensive validation, we stratified our approach into three layers: divisions, areas, and methodologies. First, we scoped four divisional targets: *software architecture, design process, arti-*

ficial intelligence, and knowledge acquisition. Then, we decomposed the divisions into

their corresponding validation areas, each having respective methodologies comprised of

employable standards, techniques, and methods (Table 5.25).

Table 5.25a. Evolving architectural standards.

Title	Date	Source	Architecture Clarifications
Architecture for KBE System-of-Systems	2006	Intelligent Systems, Inc. (Azad)	Evolving definition: Architecture for a large-scale system-of-systems project comprised of many constituent complex, distributed systems in fields such as design, systems engineering, and agent-based modeling leveraging Web Services and Semantic Web
Model-Driven Architecture	2001	Object Management Group (OMG)	Functional approach to architecture: Based on a platform-independent model (PIM) plus one or more platform-specific models (PSM) and sets of interface definitions, MDA development focuses first on the functionality and behavior of a distributed application or system, thereby divorcing implementation details from functionality
International Council for Systems Engineers (INCOSE) standard	2000	INCOSE	Definition: The fundamental and unifying system structure defined in terms of system elements, interfaces, processes, constraints, and behaviors
ANSI/IEEE Std. 1471-2000	2000	IEEE	AWG Definition: *The fundamental organization of a system, embodied in its components, their relationships to each other and the environment, and the principles governing its design and evolution.* (Recommended Practice for Architectural Description of Software-Intensive Systems)
National Defense Authorisation Act for Fiscal Year 1996 (Clinger-Cohen Act)	1996	U.S. Congress	Legal Definition: Information Technology Management Reform Act of 1996 *An integrated framework for evolving or maintaining existing technology and acquiring new information technology to achieve the agency's strategic goals and information resource management.*
First International Workshop on IT Architectures for Software Systems	1995	International IT Workshop	Categorization: 5 architectural models categorized and defined by Shaw: *Process Model, Framework Model, Structural Model, Dynamic Model, and Functional Model.*

Table 5.25b. Total validation targets divisions, areas, methodologies in AIDF-SVM.

Target	Application Areas	Validation implementation standards, techniques, methods
(I) Software Architecture	(A) Architecture development process	*(1) ANSI/IEEE Std 1471 (2) Clinger-Cohen Act (3) OMG-MDA (4) First International Workshop on IT Architectures for Software Systems in 1995, (5) Shaw-Carnegie Mellon architectural categorizations (6) ATAM for architectural trade-off analysis, (7) CBAM for architectural cost assessment, (8) Acclaro Design for Six Sigma (DFSS) implementation, (9) Telelogic TAU implementation of UML and SysML, (10) Rational Unified Process (11) Architecture-Driven Software Construction*
	(B) Architecture framework model	*(1) AIDF implementation/Acclaro Design for Six Sigma, (2) Failure Mode and Effects Analysis (FMEA), (3)Design Structure Matrix (DSM) dependency resolution, (4) Axiomatic Design Theory (ADT) risk mitigation, (5) Quality Function Deployment (QFD) (6) Generic Architecture for Upgradeable Real-Time Dependable Systems (GAURDS) validation framework (7) Survey of 25 decision support tools and 10 frameworks*
	(C) Architectural structural/dynamic model	*(1) High Level Integrated Design Environment (HIDE) for dependability (2) Telelogic TAU SysML implementation/code validation/verification*
	(D) Architectural process model	*(1) Axiomatic V-Model mapping process (2) National Academy of Engineering 2001 Report on Approaches to Engineering Design*
(II) Design Process	(A) Engineering product design phase	*Methodology by Pahl and Beitz*
	(B) Prevailing engineering design methods currently available	*(1) Axiomatic Design Theory (ADT), (2) Theory of Inventive Problem Solving (TRIZ), (3) Hierarchical Multi-layer Design (MLH), (4) Quality Function Deployment (5) Design Structure Matrix (DSM) (6) Fault-Tree-Analysis (FTA) (7) Failure-Mode and Effects Analysis (8) Reliability Block Diagram (RBD) Analysis (9)Technology Risk Factor (TRF) Assessment (10) Entropy (ETP) Analysis, and (11) Case Study: Optical Backplane Engineering (OPT) Domain*
(III) Artificial Intelligence inference Mechanisms	(A) Automated Knowledge Analysis and Design Synthesis Approach	*(1) the de-facto standard for knowledge modeling as defined by Common Knowledge Analysis and Design Synthesis (CommonKADS) Approach, and the (2) the European Union Esprit II VALID project exploring the state-of-the-art for validation of KB systems*
	(B) Knowledge-based Ontology Implementation	*(1) Concept Map tools (CmapTools) for domain expert knowledge capture with ontologies, (2) Concept Map tools (CmapTools) Methodology to connect to global knowledge repositories on Semantic Web by demonstrating the exportability of ontology constructs into XML and Web Ontology Language (OWL) for use by intelligent agents*
	(C) Knowledge-based Ontology Evaluation	Mathematical assessment of ontology quality and quantity in terms of *breadth, fan-out,* and *tangledness*
	(D) Knowledge-based Weighted Rules Verification	*(1) verification with HP Labs Jena coding for inference engine, (2) code evaluation of Java Expert System Shell (JESS), (3) weighted rules verification with Matlab Fuzzy Logic Toolbox*
	(E) Knowledge-based Algorithmic Methods	*(1) domain rule support (2) predicate logic support (3) algorithmic reasoning support (4) fuzzy logic support (5) neural network support (6) genetic algorithm support (7) Conant transmission support (8) Calibrated Bayesian Support (9) Data Mining Support*
(IV) Global Knowledge Acquisition Process	(A) Authentication of Remote Knowledge Repositories	*Cluster-on-demand architecture (Coda)*
	(B) Intelligent Agents on Semantic Web	*DAMLJESSKB and its evaluation and verification methods*

VALIDATION RATIONALE FOR EACH DIVISION

VALIDATION RATIONALE FOR SOFTWARE ARCHITECTURE

Validation Rationale for Architecture Development Process

Our validation methodology for the *architecture development* process included research on the most current architectural standards technology and global standards used by the software architecting and engineering profession today. The rationale for this architectural development validation was to assure that all of definitions used by the AIDF were compatible and consistent with those used in other parts of the world. We used the standard *ANSI/IEEE Std. 1471* because that is currently the recommended terminology for architectural views, also consistent with Shaw's architectural categorizations developed in the *First International Workshop on IT Architectures for Software Systems in 1995*. In this landmark conference, software architecture was recognized as an emerging discipline. Shaw, a leading member of the Carnegie Mellon school of thought, produced a paper distilling and categorizing the various types of architectures that were popular into five main categories called framework model, structural model, dynamic model, process model, functional model.

Soon after this conference, the U.S. congress sought a commercially viable definition of architecture by passing the *National Defense Authorisation Act for Fiscal Year 1996*, which we determined to be consistent with Shaw and *ANSI/IEEE Std. 1471*. We reviewed the state-of-the-art of the Object Management Group (OMG) which is charge of setting standards for distributed object-oriented systems, which includes a focus on modeling as well as model-based standards. The results of our research showed that the current trends of Model-Driven-Architecture (MDA) were consistent with our goal of technology independence for the AIDF.

In order to provide a methodology for a quantitative approach to architectural decision-making, we reviewed the Architectural Trade-off Analysis Method (ATAM) for alternative applications of the AIDF. Along with the ATAM, we reviewed the Cost Benefit Analysis Method (CBAM) to determine a methodology for architectural cost assessment for the AIDF. After reviewing axiomatic design theory, endorsed in the 2001 report to the nation in *Approaches to Engineering Design* by the *National Academy for Engineering*, we identified the axiomatic V-Model as our process model, as defined by Shaw, to guide the integration of the AIDF architectural framework model with the structural and dynamic models.

Validation Rationale for Architectural Framework Model

We were able to successfully implement and validate the AIDF *architectural framework* using Acclaro DFSS, a tool specialized for architectural development, recently acquired by General Dynamics for large-scale engineering projects. The fact that a company noteworthy for aeronautical engineering purchased this tool was further vindication that that the presentation of the architecture in this format would be appreciated in industry. The rationale for validation of the framework is to assure that the central focus of the thesis is validated by demonstrating front-end validation techniques. The facilities available on Acclaro DFSS directly for assisting with this validation purpose are Axiomatic Design Theory (ADT), Design Structure Matrix (DSM), Quality Function Deployment (QFD), and Failure Mode and Effects Analysis (FMEA). These techniques provided a means to validate the case study implementation by hierarchical decomposition of functional requirements to avoid implementation trap and requirements creep, resolving unnecessary module dependencies, and complexity reduction achieved by application of design axioms and theorems endorsed by the National Academy of Engineering.

Once the validation of the AIDF engine blocks are accomplished using the DSM facility of Acclaro DFSS, along with the application of the ISO/IEC 15288 and 12207 international draft standards for systems and software processes. The rationale for applying the DSM to this aspect of the architecture framework is that the analysis provides a method for stable modular expansion by eliminating dependencies, while the ISO/IEC 15288 provides a standard for software maintenance to preserve engine integrity, recognized by international organizations such as the International Council for Systems Engineers.

Once the functional and non-functional requirements of the architecture are identified and validated, as needed, we introduce a validation framework for real-time application of the KBE systems configured by the AIDF used by safety-critical systems such as nuclear submarines, space systems, and railways. This validation framework based on Generic Architecture for Upgradeable Real-Time Dependable Systems (GAURDS) provides a means to protect against real-time vulnerabilities yet to be anticipated by the AIDF architectural framework during for future expansion of its platform capability, allowing further integration of validated COTS components. As part of the research KBE systems based on the AIDF, we surveyed twenty-five decision support tools and 10 conceptual design frameworks. This enabled us to appreciate the uniqueness of the AIDF and its application to optical backplane engineering, as well as its broad impact to computer engineering.

Validation Rationale for Architectural Structural/Dynamic Model

The validation methodology for the *design environment* was accomplished with High-Level Integrated Design Environment for dependability. This approach provides an effective methodology for validation using object oriented constructs such as UML. The rationale for selecting this approach to validation and verification of the design environment is that it provides a prescription for validation that can leverage the latest development in systems engineering, specifically with SysML, since SysML extends UML.

Validation Rationale for Architectural Process Model

We used the axiomatic V-model as the process model that allows us to map the functional requirements (FR), identified in the axiomatic hierarchical decomposition process, to object-oriented methods. Showing this linkage of the product of the architectural framework to software engineering methods is important to demonstrate the practicality of this approach for both systems and software engineering.

VALIDATION RATIONALE FOR DESIGN PROCESS

Validation Rationale for the Design Process

The validation methodology for the design process was accomplished based on identification of the three primary engineering phases by Pahl and Beitz [Pahl and Beitz, 1988], widely accepted in the literature of product engineering. These are highlighted in their book *Engineering Design: A Systematic Approach.* The rationale for identification

of these phases as defined by conceptual, embodiment, and detailed design, was crucial to

developing the corresponding stages in the AIDF Design Process model that showed how

the AIDF provides automated design support during each phase.

Validation Rationale for the Design Modules

The validation methodology for the design modules was accomplished by identi-

fying 11 major engineering design theories that can provide automated design process

support based on established engineering disciplines. The rationale for selection of these

methods is that each of them is a fundamental engineering approach that is also suitable

for knowledge engineering. Thus, the modules can be developed through this knowledge

engineering process in a way that shows how to capture and convert the knowledge into a

format for machine processing. The Axiomatic Design Theory (ADT) module is based on

established engineering theory endorsed by the National Academy for Engineering and is

a theory that has been purported to be readily automated [Suh, 2001]. The Theory of In-

ventive Problem Solving module is based on an established engineering method that is

very suitable for automation due to its information-rich matrix offering inventive solu-

tions derived from millions of patent searches. Hierarchical Multilayer Design (MLH)

module is based on an engineering method extending ADT that assesses the effect and

cost of critical components to total design in a format very suitable for automation. The

Quality Function Deployment (QFD) module is based on an established engineering

methodology that can capture the voice of the customer, which can be readily automated

due to its information-rich compositional structure. The Reliability Block Diagram

(RBD) module is based on an established engineering method that can be automated due to its probability calculations requiring brute-force techniques for assessing networked component reliability. The Failure Mode and Effects Analysis (FMEA) module is based on an established engineering methodology that can be used to evaluate the potential reasons for risk and readily automated due to its information-rich structured input. The Technology Risk Factor (TRF) module is based on a methodology developed at MIT by a NASA engineer to assign weighted rules to each component indicating the reliability factor for each that multiplies into a matrix. The Entropy (ETP) module is based on an established study in physics that assesses the disorganization of the system which can be measured and reported to the user as the system complexity grows. The Optical Backplane Engineering Domain (OPT) module is based on a suitable domain that a KBE system can be applied.

VALIDATION RATIONALE FOR ARTIFICIAL INTELLIGENCE

Validation Rationale for Automated Design Synthesis and Analysis

The validation methodology for the automated design synthesis and analysis was accomplished using *Common Knowledge Analysis and Design Support (KADS),* considered as one of the state-of-the-art global standards for KBE. The rationale for selecting CommonKADS was to find a methodology for knowledge modeling that was compatible with UML and SysML, as well as MML, especially since it is based on object-oriented diagrams of the same type. This compatibility has enabled us to the modules housed in the AIDF engine block together seamlessly at the architectural level using structural and dynamic models. Furthermore, the functions defined for synthesis and analysis are in a format allowing *extensibility* to the AIDF. CommonKADS, originating in Europe, is also consistent with *Esprit II VALID project,* the three-year European Union initiative on assessing the state-of-the art of validation for KB systems.

Validation Rationale for Knowledge-Based Ontology Implementation

The validation methodology for ontology implementation of the case study was accomplished by testing a specialized visual ontology editor using Concept Map Tools, also known as CmapTools [IHMC, 2006]. We tested the ease of use and exportability for the visual editor to determine if it was capable of readily producing ontologies that could be exported into XML and OWL format. Using CmapTools we were able to construct an actual ontology for the case study describing a type of Free-Space Optical-Interconnect (FSOI) having over a dozen components. We demonstrated the scalability of this ontology generated using this tool for describing thousands of components represented as ontologies in the knowledge base. The rationale for testing CmapTools for ease of use is that we hoped to minimize the learning time needed to master a development tool by knowledge engineers, as well as domain experts responsible for developing the knowledge base ontologies. The rationale for testing CmapTools for exportability to XML and OWL was to ensure that intelligent agents that can only read machine language would be able to utilize the domain-specific knowledge developed via an ergonomic visual editor environment.

Validation Rationale for Knowledge-Based Ontology Evaluation

The validation methodology for evaluating the quality of the ontologies was de-

veloped by applying mathematical analysis to *ontology metrics*. Equations for analyzing

factors such as ontology breadth, fan-out, and tangledness were applied to assess the

quality of the ontologies. The rationale for providing a systematic evaluation methodol-

ogy allowed the development of a general schema for quality calculation that is applica-

ble to all ontologies developed for scalable KBE systems, configured and launched by the

AIDF.

Validation Rationale for Knowledge-Based Weighted Rules

The validation methodology for the *weighted rules* can be accomplished by im-

plementing the rules using the MATLAB Fuzzy Logic Toolbox, Java Expert System

Shell (JESS), and Jena of HP Labs. The code and implementations developed by these

tools can be invoked by the algorithmic processes in Tier 2, where architectural descrip-

tion at the structural and dynamic level is provided using UML/SysML. All of these tools

have recommended verification or validation facilities available. The fuzzy logic capabil-

ity of the MATLAB Toolbox can be used to develop weighted rules with a facility for

validation. JESS is a relatively new tool that can be used to test the functionality of rules

directly by invoking Java procedure calls. Jena is a Java framework for building Semantic

Web applications with an extensive array of capabilities for building a verifiable rule-

based inference engine using OWL. The rationale for recommending these tools for de-

veloping the rule-base is due to the validation and verification methods that are being introduced into these tools during their continuous development by MATLAB, Sandia National Laboratories, and HP Labs Semantic Web Program.

Validation Rationale for the Algorithmic Methods

The validation methodology of the AI modules was accomplished by identifying 9 AI techniques that can provide automated design process support based on established AI algorithms and methods. The reason for selection of these methods is that each of them is a fundamental AI approach that is also suitable for knowledge engineering. Thus, the modules can be developed through this knowledge engineering process in a way that shows how to capture and convert the knowledge into a format for machine processing. The domain rule support (DRS) module provides a framework to build a rule-base, usually in the form of unweighted IF-Then heuristic structures. The Predicate Logic Support (PLS) module provides a framework to develop inference rules based on accepted notions in computer science that provides forward-chaining and backward-chaining inference mechanisms. The Algorithmic Reasoning Support (ARS) module provides a framework for constructing rigorous algorithmic solutions. The Fuzzy Logic Support (FLS) module provides a framework to construct less rigorous weighted rules. The Neural Network Support (NNS) Module provides a framework to incorporate neural network learning based on the provided domain expertise. The Genetic Algorithm Support (GAS) module provides a framework to conduct genetic algorithm optimization searches. The Conant Transmission Support (CTS) provides a framework to develop component influence diagrams for visual display. The Calibrated Bayesian Support (CBR) module provides a framework to develop Bayesian belief networks. The Data Mining Support (DMS) module provides a framework to develop methods for data mining.

VALIDATION RATIONALE FOR GLOBAL KNOWLEDGE ACQUISITION PROCESS

Validation Rationale for Authentication of Remote Knowledge Repositories

Domain expertise stored in knowledge base, both locally and remotely, requires authentication. The Knowledge Base (KB) of the KBE SoS architected by the Artificial Intelligence Design Framework (AIDF) stores the domain based rules and ontologies supplied by the domain experts, which can be exported into XML and OWL by various ontology editors.. Considering the size and scope of the KB, it becomes inevitable to ensure that the KB conforms to high levels of security and reliability.

One way of security authentication is by ensuring that the domain experts updating are authorized. Ensuring a reliable and secure KB would involve replicating the KB averting single point failure, routine verification of the knowledge for coherence of replicated copies ensuring that all the copies are up to date, maintaining local copies for improving on access speed, providing disconnected operation. In order to achieve the above mentioned goals, an advanced next generation distributed file system such as Cluster-on-demand-architecture (Coda), a variant of Andrew File System (AFS), can be used. Both systems were developed at Carnegie Mellon University.

Validation Rationale for Semantic Web Intelligent Agents

An agent for web automation should be able to manipulate knowledge in web documents, provide different services and interact with other agents or users. To meet these requirements, a generic architecture of intelligent agents consists of *inference*, *control* and *communication* layer. The control layer is generally composed of semantic

rules and domain rules that need to be validated for accuracy, as in the case of

DAMLJESSKB retrieving marked-up components in OWL [Kopena, 2003].

SYNERGISTIC VALIDATION METHODOLOGY APPLICATION

VALIDATION METHODOLOGY FOR ARCHITECTURE DEVELOPMENT

We show the validated development process for a KBE system with AIDF-SVM (Table 5.26).

Table 5.26. Validated development process of KBE system AIDF-SVM.

AIDF-SVM-KBE-2006 Validation Process
begin: Customer expresses AI needs *informally* to develop a particular type of domain-specific KBE system, e.g. OBIT, based on the *AIDF-SVM-KBE-2006 approach to validation and verification of KBE systems*
Step 1: Customer elicitation experts, knowledge engineers, and high-level software architect captures the expressed needs *formally* for domain specific KBE system, e.g. OBIT, using AIDF architectural framework with Acclaro DFSS requirements tool in terms of hundreds of rows of functional requirements through hierarchical decomposition
Step 2: Acclaro DFSS captures explicit functional requirements (FR), e.g. descriptions of reasoning ability functional needs, producing a design matrix to optimize selection of best design parameters (DP), e.g. modules, to meet the needs of the FR, applying information and independence axiom, if possible, to reduce complexity
Step 3: System engineers examine each row of functional requirements having an assigned DP set in Tier-1 for valid functional design - iteratively – then, when validated, continue to determine the location for each of the object oriented (OO) connections to be made in detail by the software engineers in Tier-2 using Acclaro DFSS design matrix facility that shows the assigned DP set for each FR row
Step 4: After groups of software engineering teams receive the outsourced DP allocations for each FR from the systems engineers, each team is allowed to focus only on each FR row relevant to achieve the team's specified goals. This means that each team independently develops the detailed OO UML and SysML constructs defining each FR by taking into account each DP in the FR row in their OO design for that particular function, thereby preserving OMG's MDA technology independence strategy
Step 5: Systems engineers focus on smooth integration into a large-scale UML model constructed according to AIDF specifications depicting OO solutions to each FR using the allocated DPs for all the outsourced rows of FR
Step 6: Axiomatic Design V-Model provides systematic approach to integration of the UML and SysML constructs developed by hundreds of teams corresponding to each FR, which is independently developed by a team of software engineers in Tier 2
Step 7: Generic Architecture for Upgradeable Real-time Dependable Systems (GAURDS) and High Level Integrated Design Environment (HIDE) for Dependable systems are used to validate and verify architectural integrity of the KBE system
Step 8: Knowledge base validation process is continuously in process globally according to validated approach to design process model (Conceptual, Embodiment, Detailed Stages), as domain experts input and update their knowledge into their respective modules via templates and other mechanisms such as CmapTools and Fuzzy Logic weighted rule, and design structure matrix component interaction data
Step 9: Individual processing modules for AI inference and design calculation housed in the design and AI engine block, respectively, of the AIDF architectural framework are independently inserted and validated one at a time or in parallel by a combination of GAURDS, HIDE, and mechanisms for VVT for each unique module
Step 10: Coding and iterative implementation according to requirement specifications and guidelines established in tier-1 through tier-4 to ensure validation and verification of large-scale KBE system for any engineering domain
AIDF-SVM-KBE-2006 Validation Process
completion: full-scale operational KBE system is built based on AIDF architectural framework for any engineering discipline, e.g. optical backplane engineering

MULTI-TIERED IMPLEMENTATION OF THE AIDF

A landmark conference helped define categories of architecture in the *First International Workshop on IT Architectures for Software Systems in 1995.* In this conference, five architectural categories were distilled, identified, and published by Shaw, as she established herself as an authority on software architecture at Carnegie Mellon and IEEE architecture working group. Her school of thought on what constituted software architecture has survived the test of time and is widely accepted by many in the software engineering community who understand that software architecting itself is an emerging discipline from an art form. Hence, the strategic move to accept Shaw's categorization of architecture into five types was an important decision that has enabled us to more precisely define the architectural framework in terms of a functional architecture and make the approach to construction and validation of KBE systems based on the AIDF a reality through a multi-tier approach (Fig. 5.4).

In other words, when combining Shaw's approach with Suh's axiomatic Process Model, a new way to build systems is realized, in which the framework model (implemented with Acclaro DFSS) serves as Tier-1, responsible for capturing the requirements specifications of the system through a set of functional requirements and design parameters. Combined with Tier-1, the structural and dynamic models (implemented with Telelogic TAU) serves as Tier-2, responsible for capturing the object-oriented constructs that show how the functional requirements are realized in more detail.

The final stage of this three-tier approach is Tier-3 which serves as the implementation stage of the structural models as manifested in actual coding of the objects, as dictated by Suh's axiomatic approach to architectural design as expressed in the V-Model.

For instance, many of the AI and Design modules serve as the design parameters fulfilling the functional requirements in Acclaro DFSS (Tier-1).

Then, each of the functional requirements for each module serves as a guide for construction of object-oriented constructs defining detailed modular operations in a tool such as Telelogic TAU (Tier-2). Once tier 1 and tier 2 are complete, actual construction of a KBE system with coding for real-time systems for a particular application such as OBIT can begin in Tier 3.

Table. 5.27a. **Mapping tiers to AIDF models**

#	Architecture Model Type	Tier-level
1	*Process Model*	Glue
2	*Framework Model*	Tier-1
3	*Structural Model*	Tier-2
4	*Dynamic Model*	Tier-2
5	*Functional Model*	Tier-1
6	*N/A*	Tier-3
7	*GAURDS validation framework*	Tier-4
8	*High Level Integrated Design Environment (HIDE)*	Tier-4
9	N/A	Tier-5

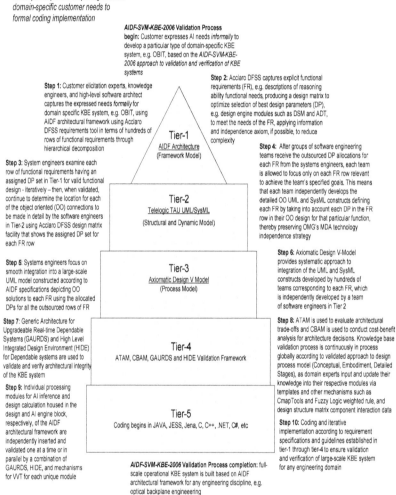

Validation methodology of a KBE system based on the AIDF-SVM-KBE-2006 approach, guiding the architectural development process capturing informal domain-specific customer needs to formal coding implementation

AIDF-SVM-KBE-2006 Validation Process

begin: Customer expresses AI needs *informally* to develop a particular type of domain-specific KBE system, e.g. OBIT, based on the *AIDF-SVM-KBE-2006 approach to validation and verification of KBE systems*

Step 1: Customer elicitation experts, knowledge engineers, and high-level software architect captures the expressed needs *formally* for domain specific KBE system, e.g. OBIT, using AIDF architectural framework using Acclaro DFSS requirements tool in terms of hundreds of rows of functional requirements through hierarchical decomposition

Step 2: Acclaro DFSS captures explicit functional requirements (FR), e.g. descriptions of reasoning ability functional needs, producing a design matrix to optimize selection of best design parameters (DP), e.g. design engine modules such as DSM and ADT, to meet the needs of the FR, applying information and independence axiom, if possible, to reduce complexity

Tier-1
AIDF Architecture
(Framework Model)

Step 3: System engineers examine each row of functional requirements having an assigned DP set in Tier-1 for valid functional design - iteratively – then, when validated, continue to determine the location for each of the object oriented (OO) connections to be made in detail by the software engineers in Tier-2 using Acclaro DFSS design matrix facility that shows the assigned DP set for each FR row

Step 4: After groups of software engineering teams receive the outsourced DP allocations for each FR from the systems engineers, each team is allowed to focus only on each FR row relevant to achieve the team's specified goals. This means that each team independently develops the detailed OO UML and SysML constructs defining each FR by taking into account each DP in the FR row in their OO design for that particular function, thereby preserving OMG's MDA technology independence strategy

Tier-2
Telelogic TAU UML/SysML
(Structural and Dynamic Model)

Step 5: Systems engineers focus on smooth integration into a large-scale UML model constructed according to AIDF specifications depicting OO solutions to each FR using the allocated DPs for all the outsourced rows of FR

Step 6: Axiomatic Design V-Model provides systematic approach to integration of the UML and SysML constructs developed by hundreds of teams corresponding to each FR, which is independently developed by a team of software engineers in Tier 2

Tier-3
Axiomatic Design V Model
(Process Model)

Step 7: Generic Architecture for Upgradeable Real-time Dependable Systems (GAURDS) and High Level Integrated Design Environment (HIDE) for Dependable systems are used to validate and verify architectural integrity of the KBE system

Step 8: ATAM is used to evaluate architectural trade-offs and CBAM is used to conduct cost-benefit analysis for architecture decisions. Knowledge base validation process is continuously in process globally according to validated approach to design process model (Conceptual, Embodiment, Detailed Stages), as domain experts input and update their knowledge into their respective modules via templates and other mechanisms such as CmapTools and Fuzzy Logic weighted rule, and design structure matrix component interaction data

Tier-4
ATAM, CBAM, GAURDS and HIDE Validation Framework

Step 9: Individual processing modules for AI inference and design calculation housed in the design and AI engine block, respectively, of the AIDF architectural framework are independently inserted and validated one at a time or in parallel by a combination of GAURDS, HIDE, and mechanisms for VVT for each unique module

Tier-5
Coding begins in JAVA, JESS, Jena, C, C++, .NET, C#, etc

Step 10: Coding and iterative implementation according to requirement specifications and guidelines established in tier-1 through tier-4 to ensure validation and verification of large-scale KBE system for any engineering domain

AIDF-SVM-KBE-2006 Validation Process completion: full-scale operational KBE system is built based on AIDF architectural framework for any engineering discipline, e.g. optical backplane engineeering

Fig. 5.4. Five-tier cascading validation approach.

MAPPING METHODOLOGY OF MODEL TO IMPLEMENTATION TOOL

In order to implement the AIDF architectural framework to build a KBE system that provides decision support for a design engineer via automated artificial intelligence, the most current tools available must be selected to match the architecture models developed. We used Acclaro DFSS to capture the high-level requirements specifications in the form of functional requirements and design parameters to produce the AIDF architectural framework that meets the needs of our particular domain-specific OBIT KBE system's expressed and evolving needs, which was combined with GAURDS validation framework to ensure we have a methodology to validate the entire system, i.e. to ensure the design of the "right" system. We used Telelogic TAU to implement the structural and dynamic models using OO approach, where each FR row in the framework model in Tier-1 provided a starting point to develop a set of objects fulfilling that particular FR's decomposition, thus the OO in Tier-2 were mapped to FR in Tier-1 using axiomatic V-Process Model (Table 5.27).

Table 5.27b. AIDF leverages architecture model types for robust description.

#	Architecture Model Type	Architectural Expression	Recommended AIDF Implementation method	Tier-level
1	*Process Model*	Framework	8-step Axiomatic Design V-Model	Glue
2	*Framework Model*	Framework	Acclaro Design for Six Sigma (DFSS)/MS Visio	Tier-1
3	*Structural Model*	ADL/OO diagrams	SysML 1.0/ UML 2.0 /MML /CommonKADS	Tier-2
4	*Dynamic Model*	ADL/OO diagrams	SysML 1.0/UML 2.0/MML SWS/Visual OWL/CmapTools/model-view-controller	Tier-2
5	*Functional Model*	Framework	Acclaro DFSS	Tier-1
6	*N/A*	N/A	coding	Tier-3
7	*GAURDS validation framework*	Architecture validation framework for Tier-1	Applying GAURDS to functional architecture of AIDF	Tier-4
8	*High Level Integrated Design Environment (HIDE)*	Architecture validation framework for Tier-2	Applying HIDE to Telelogic TAU UML/SysML diagrams	Tier-4
9	N/A	N/A	Coding of models	Tier-5

SUPPORTIVE VALIDATION RESEARCH WORK

VALIDATION TERMINOLOGY

Validation is emerging as a discipline from an art form in the field of architecture, as well as in Knowledge-Based Engineering (KBE). The process of *validation* is concerned with designing the *"right" system* that meets user expectations, whereas *verification* is concerned with designing the *system "right"* that meets engineering specifications. Hence, when a verified system's *features* correspond to its validated *benefits*, a real-world system can be developed that addresses not only rigorous functional requirements defining a system, but also stipulated application needs stated by the project sponsor, who is usually the customer requesting a domain-specific KBE system. For a more detailed account of validation definitions, the Esprit II VALID project produced a multitude of questions as well as answers as to what entailed validation for KBS [ESPRIT 2000b]. Many experts on validation have defined validation, verification, evaluation, and testing in their own words, including the prevailing standard expressed in ANSI/Sts 1471 [IEEE-1471, 2006].

BEST PRACTICES FOR DECISION SUPPORT

Twenty-Five Commercially Available Tools Surveyed

Before implementing and validating the AIDF, we conducted extensive preliminary research on architecture, artificial intelligence, and automated support for the design process for optical backplane engineering, including knowledge engineering methodologies for twenty modules. The research included a survey of twenty-five decision support and artificial intelligence tools commercially available, as part of the deliverable for a formal graduate course on Artificial Intelligence at UAB.

Fuzzy Logic and Bayesian Net Tools Demonstrated

Various rule-based and Bayesian net tools were acquired, tested, and presented for graduate-level course Artificial Intelligence as a demonstration of current technology in the areas of MATLAB Fuzzy Logic and Bayesian Nets. Two very small-scale proto-types of a fuzzy-logic controller using a weighted rule-base controlling the output rec-ommendations for a NASA Mars explorer and solar panel were demonstrated using Simulink as a deliverable for the Artificial Intelligence course at UAB.

Ten Conceptual Design Frameworks Reviewed

Dozens of frameworks for conceptual design tools and frameworks were exhau s-tively reviewed, in addition to over thirty commercial decision support systems, allowing us to conclude that the one of the AIDF architectural framework's strength is that it ad-dresses all three phases of the design process [Wang et al., 2001].

AIDF Application Uniqueness Determined

Furthermore, it was determined that the combination of the scope, broad impact, and case study application of the AIDF was truly unique. Finally, there is no evidence in the literature that a KBE system having twenty unique modules acting synergistically has ever been developed [Wang et al., 2001]. In fact, one branch of FSOI is so new that its technological infancy would greatly benefit from an AIDF KBE system design environment assisting optical backplane designers desperately needing a level of technical dexterity and precision not available by humans [Ayliffe, 1998]. Due to the potential benefits of the AIDF applied to optical backplane engineering in this particular field, laboratory FSOI applications will finally be available commercially through KBE support to its design process.

PEER-REVIEWED DEVELOPMENT PROCESS

Our process for the *peer-reviewed development* of the AIDF included consultations with the AI Group at NASA over two years, presentations at five conferences held by IEEE, SDPS, and ACM, and assimilating the feedback from referees reviewing our publications for these conferences. The reason for peer-reviewed development of the AIDF was to develop the concepts of the AIDF with feedback from authorities who could influence its architecture. For instance, we were required to report progress of our research regularly to NASA contacts in the AI group at NASA Marshall Space Flight Center. Thanks to our interactions, presentations, and feedback over a two-year period, we were able to improve on the architectural performance metrics of the AIDF. Five design conferences were attended over an 8 year period (1998-2006) that helped strengthen our understanding of the demands of an architectural framework addressing the needs of a KBE system for optical backplane engineering (SDPS, ACM, IEEE). Various journal publications to refereed journals also were valuable in getting feedback on how to improve on the AIDF, especially on methods for authenticating remote knowledge repositories on the Semantic Web.

SOFTWARE ARCHITECTURE VALIDATION

ARCHITECTURAL FRAMEWORK VALIDATION FOR KBE SYSTEM-OF-SYSTEMS

Architecture Development of System-of-Systems Application

As previously noted, System-of-Systems applications encompass many fields,

including architecting, complexity, design, and systems engineering [Madni,

2006]. Thus, in order to address an SoS problem in the realm of artificial intelligence, an

architecture framework must provide a means to achieve intellectual control over a set of

complex systems embedded in a KBE system for *design* support in *systems engineering.*

This type of automated design process support can be achieved by providing a means to

calculate for risk mitigation and reliability via design and algorithmic methods housed in

a *modular* format. As previously mentioned, Kotov's characterization of SoS as being a

large-scale, distributed, system comprised of constituent complex systems provides a ra-

tionale for a high-degree of modularization of an architecture [Kotov, 1997].

Hence, the modularization of the architecture provides an architectural foundation to

divide the functions of the large-scale system into constituent complex systems, each

having their own intricate formulations contributing to one purpose: *automated artificial*

intelligence design support.

Adaptable, Reconfigurable, Scalable Architecture for KBE System-of-Systems

In order to have a system capable of *automated artificial intelligence design*

support, we have to develop an architectural framework for a KBE System-of-Systems.

The architecture framework should be reconfigurable and scalable in order to adapt to

changing real-world needs and be versatile in its application. Contrary to KBE systems

in the past designed specifically for only one domain, a KBE system based on a *recon-*

figurable and *scalable* architecture actually thrives on rapid advances in technology and is adaptable to a spectrum of engineering challenges. In order to achieve this type of longevity and adaptation to changing needs, structural development must be achieved by such an architecture that can achieve intellectual control over a set of complex systems. A *configurable* architecture enables architecture to adapt to changing needs by adjusting its elements for an *instantiation* of a KBE system for a particular case study. An even more adaptable type, a *reconfigurable* architecture enables the KBE instantiation to be configured any number of times even after launch. This adaptability allows an entire class of KBE systems to be configured based on the same architecture, as opposed to just focusing on the idiosyncrasies of one instantiation.

Model-driven Architecture Promoting Technology Independence and Late-Binding

Another way to achieve adaptability to changing needs is by emphasizing technology independence through model-driven approach [OMG, 2006]. These types of architectures provide a means to decompose the constituent element of a given architecture into modular components that are independent of any technology, promoting the benefits of late-binding and longevity of the architecture for any application, including those for KBE systems.

Modular Architecture for systematic and concurrent development

Some of these complex systems connected to a *modular* framework are *agent-based* and must intelligently interact with both local and remote knowledge repositories automatically. One of the best ways to achieve this is by divide-and-conquer approaches through hierarchical decomposition that provides for modular architecture development. Thus, the development of the individual modules for such a modular system can proceed *systematically* and *concurrently* on different time-scales by different teams. Thus decoupling of the development processes is achieved through this modular approach for application development. Furthermore, distributed knowledge repositories being developed by domain experts in the field can also proceed methodically module by module, matching the structure of the architecture. These types of linkages requiring precise orchestration and authentication of the activity of thousands of remote domain experts updating knowledge, and we believe that intelligent agents retrieving this knowledge can be developed within a system-of-systems paradigm. An architectural framework that fulfills this need should also be able to provide an overarching structure for orderly development and integration of complex systems, including but not limited to architecture development, design process, artificial intelligence mechanisms, and agent-based knowledge acquisition techniques.

DEVELOPING QUALITY ARCHITECTURE FOR SYSTEM OF SYSTEMS

Need exists for Quality in Architecture Development

Clearly there is a need for quality in architecture development, considering more than 50% of software projects fail to meet their initial objectives [Loucopulos, 2004]. In this section we will discuss various methods for achieving real-world objectives, by ensuring the architecture quality methods provide a foundation to develop a front-end validation and verification plan for an architecture framework. When developing the right architecture for a system-of-systems application, care should be taken to ensure that intel-

307

lectual control, as well as validation, can be achieved over a wide array of complex, interdisciplinary systems acting together for one purpose. In this thesis, our focus is developing an architectural framework for a system-of-systems application.

Front-end Validation Critical for Architectural Framework

In order to provide structure for complex systems needing reliability, the development of a high-level architecture using front-end validation techniques is required. Architecture-driven software development based on model-driven-architecture and modular techniques provides facilitates development of a technology independent platform with the benefits of late-binding. An ideal architecture is developed with front-end validation techniques to last much longer than the technology that it houses, due to rapid development rate of many software technologies observed in industry. These techniques include Design for Six Sigma methods such as DSM, QFD, and FMEA.

Forward Engineering Recommended for Large-scale Projects

Historically, it has commonly been observed that software developers have tried to solve a problem with little thought on structure at the outset, resulting in project failure and/or budget crises [Maciaszek, 2001]. Large-scale projects can be divided into two main approaches, namely reverse engineering and forward engineering [Jett, 2006]. The former approach to development clearly has emerged as the dominant methodology for systematic development of software engineering projects, as observed through the development of the field of software architecting as a discipline in itself [Maier and Rechtin, 2000]. In *reverse engineering*, attempts are made to adapt software already written to pre-existing architecture, which is not a recommended approach to a large-scale system of systems project [DeLaurentis and Callaway, 2004] . In *forward engineering*, architecture development and requirements elicitation provides a foundation for front-end validation, long before any actual development takes place [Suh, 2001]. Through the forward process, object dependencies in software design can be proactively reduced in early stages by developing a structure for programmers to follow in order to produce high-quality software faster and within budget - and most importantly - that actually meets the original real-world needs of the end-user.

Quality Function Deployment based on ISO/IEC 9126 Architecture Standard

As a method of architecture evaluation, Quality Function Deployment (QFD) has played a prominent role across government, industry, and academia. It is important to verify and validate if the developed architecture actually meets the real-world needs. QFD is one of many ways to assess the actual relevance of the architectural features to the needs of a domain-specific application. In order to identify and evaluate the features with the tangible needs, a standard validation framework reference defining a model for evaluating software is ISO/IEC 9126 standard. This standard has been used as a basis to evaluate software products in terms of quality characteristics by providing information technology (IT) guidelines for their use since 1991. This standard is structured based on six primary product characteristics: *Functionality, Reliability, Usability, Efficiency, Maintainability, and Portability*. Although these six items are sufficient for a high-level design, specific technical requirements can be addressed by applying the associated sub-characteristics (Table 5.28). Expansion of this topic will be provided in the architecture development section of this chapter and its corresponding sections in later chapters.

Table 5.28. Quality function deployment of real world-needs using ISO/IEC 9126.

ISO/IEC 9126 Characteristic	Sub-characteristics
Functionality	Suitability, Accurateness, Interoperability, Accurateness, Compliance, Security
Reliability	Maturity, Fault Tolerance, Recoverability
Usability	Understandability, Learnability, Operability
Efficiency	Time Behavior, Resource Behavior
Maintainability	Analyzability, Changeability, Stability, Testability
Portability	Adaptability, Installability, Conformance, Replaceability

Assessing Impact of Architecture on Real-World Needs

In Six Sigma, QFD can be adapted to assessing the impact of an architectural framework on the real-world needs of a particular application by leveraging the ISO/IEC 9126 standard [Zrymiak, 2006]. The requirements elicitation is based on real-world needs which are then compared, parameter to parameter, with the design features provided by the system being developed. This approach can serve as a basis for developing a validation and verification plan to make sure the right architecture for the system-of-systems is built (Fig 5.5).

Applying ISO/IEC 9126 standard to Quality Function Deployment for assessing software
architecture impact on real-world needs as defined by domain-specific design features

	List of Design Features of software system (i.e. KBE) in terms of architectural, OO, and domain-specific considerations	
Checklist of real-world needs addressed by **architecture framework** based on ISO/IEC 9126	Parameter Mesh	Comparison of Real-World Priorities
	Benchmarked Target Values	

Fig. 5.5. Adapting QFD for assessing impact of architecture to real-world needs.

Additional Architecture Considerations for System-of-Systems

In order to develop a successful System-of-Systems architecture for complex systems, additional considerations should be made to standard architectural characteristics
defined by ISO/IEC 9126 standard [ISO, 2002]. These include robust design, emergence,
evolution, sustainability, and versatility are just to name a few [DeLaurentis, 2005]. Issues addressing *robust design* combine sampling-based and simulation-based perspectives to produce limited outcome probability distributions that estimate likely performance in order to assist architectural decision-makers. Issues addressing *emergence* provide a means to scan for potential pitfalls and identify beneficial trends over a period of
time, as opposed to simply attempting to assess one system at a time. Issues addressing
evolution provide a means enable the architecture to adapt to its environment over time.
Issues addressing *sustainability* provide a means to keep architectures relevant over time
in terms of effectiveness, affordability, safety, and reliability. Issues addressing *versatility*
enable the architect to maintain robustness in the presence of uncertain requirements.
Many other characteristics can be identified depending on the application (Table 5.29).

Table 5.29. System-of-Systems architecture considerations for complex system.

SoS Architecting	Key Emphasis
robust design	Produce limited outcome probability distributions estimating performance
emergence	Scan for potential pitfalls and identify beneficial trends over a period of time, as opposed to simply attempting to assess one system at a time.
evolution	Enable the architecture to adapt to its environment over time
sustainability	Keep architectures relevant over time in terms of effectiveness, affordability, safety, and reliability.
versatility	Maintain robustness in the presence of uncertain requirements.
Domain-specific Considerations	Many other characteristics can be identified depending on the application.

Six Sigma Validation Tool

Axiomatic Design reduces development risk, cost and time by providing a phased and function based structure for the architectural design process that can be combined with methods such as quality function deployment of Design for Six Sigma (DFSS) for validation purposes, especially in the Process Variable Domain [Slocum, 2006]. An architectural development tool endorsed by the founder of Axiomatic Design Theory [Suh, 2001] and recently acquired by General Dynamics [Axiomatic Design, 2005], the Acclaro DFSS tool provides facilities for front-end validation of the architectural framework before the build phase, prior to actual software development, using DSM for dependency resolution, ADT for functional decomposition and complexity reduction with axioms, TRIZ for trade-off analysis, FMEA for identification of potential problems and solutions, in addition to QFD for traceability and mapping to real-world requirements [Axiomatic Design, 2005]. Thus, complete traceability of requirements for a particular domain-specific configuration of the architecture, such as a KBE system for optical backplane engineering, can be mapped to real-world needs to achieve front-end validation before computer-aided-design (CAD) begins. More on validation will be provided in the section on validation in this chapter and the chapter devoted to a comprehensive validation methodology.

Adaptability based on Reconfigurable and Scalable Architecture

One of the important features for quality architecture is its ability to adapt to a rapidly changing technology and economic environment. This is achieved by ensuring that the architecture developed is both *reconfigurable* and *scalable*. A reconfigurable architecture is achieved when the architectural features of a KBE system can be adjusted, or configured, before and *after* development and deployment. A scalable architecture is achieved when portions of the architecture can be expanded and contracted depending on the needs of a particular KBE system. Adapting and scaling the architecture for a particular case study instantiation of a KBE system can be accomplished using techniques implemented by Acclaro DFSS, such as ADT, DSM, and QFD, as described previously.

FRONT-END VALIDATION OF FUNCTIONAL ARCHITECTURE

ARCHITECTURAL FRAMEWORK IMPLEMENTATION FOR VALIDATION

In order to validate the framework, we acquired Acclaro Design for Six Sigma (DFSS) and implemented the case study with this industrial-grade tool used by companies such as General Dynamics for architectural development and front-end validation. The auxiliary facilities of this architecture development tool enable application of Axiomatic Design Theory (ADT), Design Structure Matrix (DSM), Failure-Mode and Effects

Analysis (FMEA), and Quality Function Deployment (QFD) to enforce framework vali-

dation and six sigma quality. Acclaro DFSS provided a means for configuration of the

AIDF platform in order to develop a class of KBE systems supporting the product design

process, in particular for the case study on optical backplane engineering. The following

sections will provide more details on how this validation was achieved by implementa-

tion.

NATIONAL ACADEMY OF ENGINEERING STUDY ON STATE-OF-THE-ART DESIGN

The *National Research Council* study Improving Engineering Design [NRC,

2006] reported that the best engineering practices were not widely used in U.S. industry.

After finding that its recommendations were not implemented, the *National Academy of*

Engineering produced a report *Approaches to Improve Engineering Design* that provided

a state-of-the-art review of decision-making in engineering design, consulting with a

cross-section of engineering design leaders in industry and academia. This report has

been reviewed by individuals chosen for their diverse perspectives and technical exper-

tise, in accordance with the procedures approved by the NRC's Report Review Commit-

tee. The purpose of the independent review is to provide candid and critical comments

that will assist the authors and the NRC in making the published report as sound as possi-

ble and to ensure that the report meets the institutional standards for objectivity, evi-

dence, and responsiveness to the study charge. In this report, Suh's axiomatic design

theory was endorsed, receiving the highest rating for "selection among alternative con-

cepts". The fact that ADT is endorsed by NRC is important, considering that the Acclaro DFSS tool we used to implement the AIDF enables validation based on axiomatic design theory, among other capabilities for verification such as design structure matrix.

IMPORTANCE OF FRONT-END VALIDATION BEFORE SYSTEM DEPLOYMENT

Our aim is to validate the architectural framework *model* of the AIDF that functions as a platform for launching a class of KBE systems specialized for design support improving on speed, time, and cost parameters for product design. This front-end validation approach is intended to eliminate critical faults in the launching platform from propagating to system level, *before KBE system deployment.* Thus, risk mitigation for KBE system design is achieved at the architecture level instead of waiting for post-deployment problems commonly experienced by corporations during large-scale system deployment, evaluation and testing, which is outside the scope of this thesis.

More importantly, various front-end validation techniques are applied to the AIDF to prevent *implementation trap*, a condition where original functional requirements are not addressed or lost during downstream design decisions made during software engineering implementation [Albin, 2003]. Furthermore, *requirements creep,* a condition of late-stage functional add-ons defining KBE system needs, can not only be avoided, but addressed flexibly due to the configurability of the AIDF for each domain of application. The high-configurability of the AIDF is attributed to the modular framework structure, allowing as many modules to be added for design and AI support in the engine block; in this thesis, we address twenty unique modules that can be knowledge engineered for inductive and deductive reasoning, as well as brute-force calculations.

FRAMEWORK IMPLEMENTATION WITH ACCLARO DESIGN FOR SIX SIGMA

The AIDF architectural framework itself is implemented with the architecture requirements analysis tool *Acclaro Design for Six Sigma (DFSS)*, an architecture development tool supporting Six Sigma methods used by engineering companies, such as General Dynamics, for large-scale design projects. Supporting aspects of Six Sigma design processes for front-end validation of architecture quality, Acclaro DFSS enabled us to validate the AIDF based on Axiomatic Design Theory (ADT), Design Structure Matrix (DSM), Quality Function Deployment (QFD), and Failure Mode and Effects Analysis (FMEA). In this section, we will briefly describe *what* each technique enabled us to accomplish for the purpose of AIDF implementation validation. In the next chapter, we will describe, in more detail, *how* we accomplished these validation techniques for the case study configuration using these techniques.

Risk mitigation can be accomplished by applying the validation techniques available to the Acclaro DFSS tool. We applied the decomposition methods and design axioms, theorems, and corollaries of Axiomatic Design Theory as a means to validate the functional requirements and design parameters of the AIDF architectural framework. The ADT methodology allowed us to systematically capture, maintain, and modify, as needed, the comprehensive array of AIDF architectural functional requirements (FR) by a process of hierarchical decomposition, as well as a mapping process to one or more design parameters fulfilling the needs of these FRs. In addition to this decomposition method supported by Simon in Sciences of the Artificial, the ADT facility of Acclaro DFSS allowed us the option to apply the Independence Axiom to eliminate coupling and the Information Axiom to identify best candidate designs from a set of alternatives, evaluated based on a method to calculate information content of the design parameters reflecting design complexity. Once the AIDF design matrix array was decomposed to the leaf level, we applied the analysis methodology of *Design Structure Matrix (DSM)* to validate the architectural design efficiency by eliminating as many design dependencies between system components as possible, which we show in the case study for specific modules in the AIDF engine block. In order to make sure that the AIDF met the quality goals of our architecture in terms of cost, timeliness, and speed, *Quality Function Deployment* was applied to the AIDF architectural integrity for validation based on the ISO 9126 standard, structured along six primary product characteristics: Functionality, Reliability, Usability, Efficiency, Maintainability, and Portability. We applied the *Failure Mode and Effects Analysis (FMEA)* facility of Acclaro DFSS, as part of the Six Sigma validation process, to assess the configured KBE system's predicted performance, failure modes, reliability, and risks as a proactive technique for identification and prevention of potential KBE faults before committing time and resources to costly software development, launch, and deployment.

VALIDATION OF SOFTWARE ARCHITECTURE

Primary Validation Target is the Software Architecture

Each validation division target can be decomposed into areas, each having respective methodologies comprised of employable standards, techniques, and methods. Hence, for the sake of comprehensive validation, we divided the comprehensive approach to validation by scoping four fundamental targets: *(I) Software Architecture, (II) Design Process, (III) Artificial Intelligence Inference Mechanisms,* and *(IV) Global Knowledge*

Acquisition Process. Although much work on validation is done addressing as many areas as possible associated with the AIDF, the primary validation target is the software architecture, since the artificial intelligence design framework itself is an architectural framework. Thus, validation of the implementation concentrates primarily on the framework. However, for comprehensiveness, the other three validation targets areas are also addressed in order to develop a synergistic methodology to configure a class of KBE systems having AI and Semantic Web elements based on the AIDF architectural framework platform (Fig. 5.6).

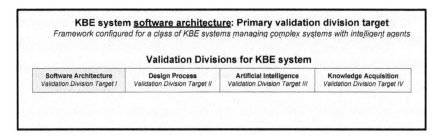

Fig. 5.6. Software architecture division is primary validation target.

Pre-deployment Validation Strengthened by Post-deployment Validation

We have separated our overall software architecture validation strategy into two stages: (1) Pre-deployment and (2) Post-deployment of a KBE System-of-System (SoS). In pre-deployment we concentrate our efforts on front-end validation using Design for Six Sigma IDOV methodology. In post-deployment, we recommend the post-deployment application of a real-time architectural validation strategy for full-scale, continuous, and industry-grade validation of the front-end validated system using Generic Architecture for Real-time Upgradeable Dependable Systems (GAURDS) validation framework (Fig. 5.7).

Software architecture validation strategy for KBE system	
KBE System ***Pre-deployment***	**Design for Six Sigma IDOV Methodology** *Front-end Architectural Validation Approach*
KBE System ***Post-deployment***	**GAURDS Validation Framework** *Real-time Architectural Validation Approach*

Fig. 5.7. Pre-deployment and Post-deployment validation strategy.

Expanding on Design for Six Sigma contrasted with Traditional Six Sigma

When applying Design for Six Sigma (DFSS), a distinction should be made for its

application to two different cases (1) already existing, or (2) to an unprecedented product

or process. If it already exists but needs improvement, the DMAIC methodology, which

stands for *Define, Measure, Analyze, Improve, control* is recommended in DFSS.

DMAIC Six Sigma usually occurs *after* initial system or product design has been

completed. However, if the product has never been developed before and needs quality

control and front-end validation, the IDOV methodology is a recommended approach for

reliability engineering, which stands for I*dentify, Design, Optimize, and Verify*.

Contrasted with this is the traditional DMAIC Six Sigma process, as it is usually

practiced (Table 5.30).

Table 5.30. Design for Six Sigma IDOV methodology for front-end validation.

Step	Purpose	Method
Identify	Improve architecture by ensuring that functional requirements actually meet the original *end-user* needs	Identify real-world needs using method such as Quality Function Deployment (QFD) based on ISO/IEC 9126 defining standard features for end-user
Design	Improve architecture by ensuring that proper risk mitigation and reliability engineering methods have been applied in the modules	Design for systems level and component level risk mitigation by considering techniques like probability risk assessment (PRA) with Fault-Tree Analysis (FTA) and Reliability Block Diagram (RBD), reliability engineering using methods such as axiomatic design theory (ADT), Failure Mode and Effects Analysis (FMEA)
Optimize	Improve architecture by ensuring that efficiency is introduced into design	Optimize the configuration to ensure that dependencies are resolve using method such as Design Structure Matrix (DSM) and Theory of Inventive Problem Solving (TRIZ)
Verify	Improve architecture by ensuring that the architected solution actually meets the original *engineering* needs	Verify that all the stated functional requirements are actually being met explicitly by the design parameters of the system by applying methods such as Axiomatic Design Theory (ADT)

GAURDS Validation Framework

The GUARDS validation strategy considers both short-term and long-term objectives, which is the validation of the design principles of the architecture and the validation of instances of the architecture for specific requirements, respectively. Many Different methods, techniques and tools have contributed to these validation objectives. The validation environment that supports the strategy is depicted in the GAURDS diagram (Fig 5.8), which illustrates also the relationship between the components and their interactions with the architecture development environment. The main validation components in GAURDS depicted in the diagram are: (1) formal verification, (2) model-based evaluation, (3) fault injection, and (4) the methodology and the supporting toolset for schedulability analysis. In the case of fault injection, which is carried out on KBE prototypes,

complements the other validation components by providing means for assessing the va-

lidity of the necessary assumptions made by the formal verification task. Also, it I useful

in estimating the coverage parameters included in the analytical models for dependability

evaluation (Fig. 5.8).

**AIDF Generic Architecture for Real-time Upgradeable Dependable Systems
(GAURDS) Validation Framework**

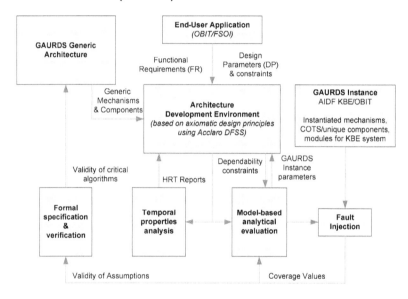

Fig. 5.8. GAURDS validation framework for KBE SoS instance.

Summary for AIDF-SVM for Software Architecture Division

In order to validate the primary validation target (Table 5.31) of the AIDF involv-

ing *Software Architecture,* four areas were identified that addressed the (A) *architecture development process*, (B) *architecture framework model*, (C) *architectural struc-*

tural/dynamic model, (D) and *architectural process model*.

Table 5.31. Software architecture validation: Standards, techniques, methods.

Division	Application Areas	Validation Methodology standards, techniques, and methods
(I) Software Architecture	(A) Architecture development process	*(1) ANSI/IEEE Std 1471 (2) Clinger-Cohen Act (3) OMG-MDA (4) First International Workshop on IT Architectures for Software Systems in 1995, (5) Shaw-Carnegie Mellon architectural categorizations (6) ATAM for architectural trade-off analysis, (7) CBAM for architectural cost assessment, (8) Acclaro Design for Six Sigma (DFSS) implementation, (9) Telelogic TAU implementation of UML and SysML, (10) Rational Unified Process (11) Architecture-Driven Software Construction*
	(B) Architecture framework model	*(1) AIDF implementation/Acclaro Design for Six Sigma, (2) Failure Mode and Effects Analysis (FMEA), (3)Design Structure Matrix (DSM) dependency resolution, (4) Axiomatic Design Theory (ADT) risk mitigation, (5) Quality Function Deployment (QFD) (6) Generic Architecture for Upgradeable Real-Time Dependable Systems (GAURDS) validation framework (7) Survey of 25 decision support tools and 10 frameworks*
	(C) Architectural structural/dynamic model	*(1) High Level Integrated Design Environment (HIDE) for dependability* *(2) Telelogic TAU SysML implementation/code validation/verification*
	(D) Architectural process model	*(1) Axiomatic V-Model mapping process* *(2) National Academy of Engineering 2001 Report on Approaches to Engineering Design*

SUMMARY

In this chapter, we have introduced the state-of-the-art for architecture, including a treatise on design process automation with KBE techniques and comprehensive validation methodology in terms of standards, techniques, and methods. In this thesis, we develop a validated AIDF architectural framework that functions as a platform for design process automation based on Semantic Web-enabled artificial intelligence. In order to achieve comprehensive validation, we recognized the need to research the state-of-the-art standards, techniques, and methods employable by KBE systems configured by the

AIDF. For comprehensive validation, we stratified our approach into three layers: divisions corresponding to multiple areas and an itemized list of methodologies.

CHAPTER VI: CONCLUSIONS, CONTRIBUTIONS AND FUTURE WORK

OVERVIEW

In this chapter, we introduce the contributions of the AIDF in terms of broad impact of the framework on design risk mitigation for reliability engineering in the optical backplane domain, as well in other fields impacted by architecture development, system-of-systems, and computer engineering. Future work discussing how to systematically expand the architectural framework scope and its possible impact on commercialization efforts in optical backplane engineering is introduced.

CONTRIBUTIONS

We have developed a framework for architecture-driven software engineering of large-scale systems that fills a niche between traditional requirements and design tools impacting three broad fields: architecture development, system-of-systems, and computer engineering. In government, a strategic need recognized by NASA for design risk mitigation can be addressed by applying the AIDF architecture-driven development process to large-scale, complex, software systems leveraging Web Services. The AIDF-SVM provides a reconfigurable and scaleable platform to develop class of design support systems that are optimized, verified, and validated to engineering specifications that can be traced directly to original functional requirements to fulfill end-user needs.

We have accomplished this design risk mitigation by using a System-of-Systems (SoS) approach, which consists of achieving intellectual control over a synergistic overlap of many broad topics such as design, agent-modeling, and systems engineering using an architectural framework that has a strategy for validation before and after deployment. Subsequent to front-end validation of the architecture framework with DFSS IDOV techniques, we provide an approach to ensconce the pre-validated AIDF into the Generic Architecture for Upgradeable Real-time Dependable Systems (GAURDS) validation framework for continuous validation after deployment, also used by safety critical systems such as nuclear submarines, space systems, and railways. This robust approach to validation provides protection against requirements creep and implementation trap as new modules are added, resulting in more executable functional requirements and design parameters expanding the capability of the AIDF.

In addition to architecture validation, we provide a comprehensive validation approach for a KBE SoS using the Synergistic Validation Methodology (SVM) for the AIDF, which provides recommendations on the types of validation methods available for architecture driven applications of a KBE SoS. The AIDF-SVM is stratified into four major target validation divisions, which are expanded into multiple areas, and mapped to various methodologies, defined in terms of standards, techniques, and methods that can serve as a guide for continuous validation before and after system deployment. The four divisions are categorized as: (1) Software architecture, (2) Design process (3) Artificial intelligence, and (4) Knowledge acquisition.

The AIDF, together with its SVM, provides software developers a systematic approach to building high-quality, low-cost KBE SoS applications, with optimized and verified expediency, for reliable product engineering that are networked to Web Services. Thus, the AIDF-SVM approach has a broad impact on software development of a class of engineering design support applications requiring automated design risk mitigation for reliability engineering, particularly for configuring optical backplane engineering, as well as for any other type of product engineering need, in general.

FUTURE WORK

The AIDF functional requirements can be expanded from 300 FR and DP nodes to thousands with hierarchical decomposition to further specify the needs of a model-driven architecture that is independent of current technology. Thus, a futuristic system can be defined before the actual modules are available, inspiring the next generation of developers to fill the void, in terms of a design fulfilling an end-user need that has yet to be developed. By applying the validation techniques in the AIDF-SVM, software developers can be assured that the system developed over time will be able to adapt to changing market conditions and government needs. This flexibility can be achieve by leveraging the advancing technology available to Web Services, as the Internet2 and the Next Generation Internet become significant sources of knowledge engineering development centers.

In the area of optical backplane engineering, the AIDF can serve as a catalyst for the commercialization of free-space applications for optical interconnect technology, in particular, while providing a configurable and scalable springboard for automating many other reliability engineering applications in other disciplines. A case study implementation of the AIDF that was developed using Acclaro Design for Six Sigma (DFSS) specifically provides FSOI engineers, developing sensitive optical equipment operating at the micron level, a new type of design support platform that leverages the Semantic Web and

at least twenty other possible design and inference support modules for automated re a-
soning during the design process. A fully-implemented KBE SoS application developed
in the future for this particular optical backplane engineering area is expected to signif i-
cantly improve the chances of success for this particular branch of optical backplane e n-
gineering, enabling it to leave the laboratory and become commercially viable.

LIST OF REFERENCES

Abdullah, M. S., A. Evans, I. Benest, and C. Kimble, 2004, "Developing a UML Profile for Modelling Knowledge-Based Systems." [Online]. Available: http://www-users.cs.york.ac.uk/~kimble/research/UML_Profile_for_Modelling_KBS.pdf

Albin, A., 2003, *The Art of Software Architecture*, John Wiley & Sons Inc., New York.

Altshuller, G., 1997, *40 principles: TRIZ Keys to Technical Innovation,* Lev Shulyak and Steven Rodman (translation), Technical Innovation Center, Worchester, MA.

Angele, J., D. Fensel, D. Landes, and R. Studer, 1998, "Developing Knowledge -Based Systems with MIKE," *Journal of Automated Software Engineering*, Vol. 5. No. 4, pp. 389-418.

ANSI standard, 2006, [Online]. Available: http://standards.ieee.org/reading/ieee/std_public/description/se/1471-2000_desc.html

Axiomatic Design, 2005, "Slideshow on Axiomatic Design and Acclaro software," [Online].Available: http://www.vr.clemson.edu/credo/classes/lect6-15.pdf#search='axiomatic%20design

Ayliffe, M.H., 1998, "Optomechanical, electrical and thermal packaging of large 2D optoelectronic device arrays for free-space optical interconnects," *SPIE*, Vol. 3490, pp. 502-505.

Ayliffe, M.H. 2001, *Alignment and packaging techniques for two dimensional free-space optical interconnects,* Ph.D. dissertation, McGill University, Montreal, Canada.

Ayliffe, M.H., D.R. Rolston, A.E.L. Chuah, E.Bernier, F.S.J. Michael, D. Kabal, A.G. Kirk, and D.V. Plant, 2001, "Design and Testing of a Kinematic Package Supporting a 32 32 Array of GaAs MQW Modulators Flip-Chip Bonded to a CMOS Chip," *IEEE Journal of lightwave technology*, Vol. 19, No. 10, pp. 1543-1559.

Ball, L.J., N.J. Lambell, T.C. Ormerod, S. Slavin, and J.A. Mariani, 2001, "Representing design rationale to support innovative design re-use: A minimalist approach," *Journal of Automation in Construction*, Vol. 10, No. 6, pp. 663-674.

Bar-Yam, Y., 2003, "When systems engineering fails -toward complex systems engineering," *IEEE International Conference on Systems, Man and Cybernetics,* Vol. 2, pp. 2021-2028.

Basilio, R.R., K. S. Plourde,. and Lam, T., 2000, "A Systematic Risk Management Approach Employed on the CloudSat Project," Jet Propulsion Laboratory, [Online]. Available: http://trs-new.jpl.nasa.gov/dspace/bitstream/2014/16290/1/00-2257.pdf

Bass, L., P. Clements, and R. Kazman, 2003, *Software Architecture in Practice,* Addison-Wesley, Reading, MA.

Bearden, D. A., 2000, "A Complexity-Based Risk Assessment of Low-Cost Planetary Missions: When is a Mission Too Fast and Too Cheap?" in Complexity based cost estimating relationships for space systems, Filippazzo, G., *IEEE Aerospace Conference, Proceedings,* 2004, Vol. 6, pp. 4093- 4098.

Beckett, J., 2005, "HP Labs Semantic Web Research," [Online]. Available: http://www.hpl.hp.com/semweb/

Bielawski, L. and R. Lewand, 1991, *Intelligent Systems Design, Integrating Expert Systems, Hypermedia, & Database Technologies*, John Wiley & Sons Inc., New York.

Boehm, B., 1988, "A Spiral Model of Software Development and Enhancement," *IEEE Computer,* Vol. 21, No. 5, pp. 61-72.

Boehm, B., 2006, "Some Future Trends and Implications for Systems and Software Engineering Processes." *Systems Engineering*, Vol. 9, No. 1, [Online]. Available: http://sunset.usc.edu/publications/TECHRPTS/2006/usccse2006-603/usccse2006-603.pdf

Booch, G., J. Rumbaugh, and I. Jacobson, 1999, *Unified Modeling Language-User's Guide*, Addison-Wesley, Reading, MA.

Brady, T. K., 2002, *Utilization of Dependency Structure Matrix Analysis to Assess Implementation of NASA's Complex Technical Projects,* Master of Science Thesis in Engineering and Management, MIT, Cambridge, MA.

Brandish M., M. Hague, and A. Taleb-Bendiab, 1996, "M-LAP: A Machine Learning Apprentice Agent for Computer Supported Design," [Online]. Available: http://web.cs.wpi.edu/~dcb/Papers/AID00-janet.pdf

Brice, A. and B. Johns, 1998, "Improving process design by improving the design process," [Online]. Available: http://www.enviros.com/drama

Broadbandrank, 2004, "High Speed Internet Service Delivery: Backplane Definition," [Online]. Available: http://www.broadbandrank.com/glossary/Backplane.html

Brown, D. C. and R. Bansal, 1991, "Using Design History Systems for Technology Transfer," in *Computer Aided Cooperative Product Development*, D. Sriram, R.

Logcher, and S. Fukuda, (eds.), Lecture Notes, No. 492, Springer-Verlag, New York, pp. 544-559.

Calvano C. N. and P. John, 2004, "Systems engineering in an age of complexity," *Systems engineering,* Vol. 7, No. 1, pp. 25-34.

Cassidy, P., 2004, "Definition: expert system," [Online]. Available: http://www.webster-dictionary.org/definition/expert%20system

Chatzigeorgiou, A. and G. Stephanides, 2003, "Entropy as a Measure of Object-Oriented Design Quality," 1st Balkan Conference on Informatics BCI'2003, Thessaloniki, Greece, November 21-23. [Online]. Available: http://eos.uom.gr/~achat/papers.html

Chen, A., B. McGinnis, D. Ullman, and T. Dietterich, 1990, "Design History Knowledge Representation and Its Basic Computer Implementation," in *The 2nd International Conference on Design Theory and Methodology*, ASME, Chicago, IL, pp. 175-185.

Clinger-Cohen Act, 1999, [Online]. Available:http://www.ed.gov/policy/gen/leg/cca.html

CODA, 2006, [Online]. Available: http://www.cs.cmu.edu/afs/cs/project/coda-www/ResearchWebPages/docs-coda.html

Concept Map, 2006, "Cmap Tools," [Online]. Available: http://cmap.ihmc.us/

Conant, R. C., 1972, "Detecting subsystems of a complex system," *IEEE Trans. Syst., Man. & Cybern.,* Vol. 2, pp. 550-553.

Corning, P.A, 1995, "Synergy and Self-organization in the Evolution of Complex Systems, "Systems *Research,* Vol. 12, No. 2, pp. 89-121.

CLIPS, 2006, "Reference Manual," [Online]. Available: http://www.ghg.net/clips/download/documentation/bpg.pdf

Conklin, J. and K. Burgess-Yakemovic, 1995, "A Process-Oriented Approach to Design Rationale," in *Design Rationale Concepts, Techniques, and Use*, T. Moran and J. Carroll, (eds), Lawrence Erlbaum Associates, Mahwah, NJ, pp. 293-428.

CRMES, 2006, "Center for Risk Management of Engineering Systems, [Online]. Available: http://www.sys.virginia.edu/sieds06/papers/FMorningSession6.4.pdf

Daconta, M. C., L. J. Obrst, and K. T. Smith, 2003, *The Semantic Web: A Guide to the Future of XML, Web Services, and Knowledge Management,* John Wiley & Sons Inc., New York.

330

DEFSTAN, 2006, Ministry of Defense, Defense Standard 00-35, Issue 4 [Online]. Available: http://www.dstan.mod.uk/data/00/035/05000400.pdf

DeLaurentis, D. A., 2005, "Understanding Transportation as a System of Systems Design Problem," 43rd AIAA Aerospace Sciences Meeting and Exhibit, Reno, Nevada, AIAA-2005-0123. [Online]. Available:
http://www.aiaa.org/content.cfm?pageid=406&gTable=Paper&gID=25161

DeLaurentis, D. A., R. K. Callaway, 2004, "A System-of Systems Perspective for Future Public Policy," *Review of Policy Research*, Vol. 21, No. 6, pp. 829-837.

DFSS, 2006, "Design for Six Sigma," [Online]. Available:
http://www.sixsigmaspc.com/dictionary/DFSS-designsixsigma.html

DOD-STD-2167A/498, 1995. [Online]. Available:
http://cost.jsc.nasa.gov/PCEHHTML/pceh.htm

Earl, C., J. Johnson, and C. M. Eckert, 2004, "Complexity," in *Design Process Improvement*. P. J. Clarkson and C. M. Eckert, Springer Verlag, New York.

El-Haik, B., 2004, *Axiomatic Quality: Integrating Axiomatic Design with Six-Sigma, Reliability, and Quality Engineering*, John Wiley & Sons Inc., New York.

Ertas, A. and J. C. Jones, 1996, *The Engineering Design Process*, 2nd Edition, John Wiley & Sons Inc., New York.

Esener, S. and P. Marchand, 2000, "Present and Future Needs of Free-Space Optical Interconnects," [Online]. Available: http://ipdps.cc.gatech.edu/2000/wocs/18001109.pdf

ESPRIT, 2000a, "Moka Language," [Online]. Available:
http://www.epistemics.co.uk/Notes/146-0-0.htm

ESPRIT, 2000b, "The EU information technologies programme," [Online]. Available:
http://cordis.europa.eu/esprit/home.html

Felfernig A., G. Friedrich, D. Jannach, and M. Zanker, 2000, "Generating product configuration Knowledge bases from precise domain extended UML models," in *12th International Conference on Software Engineering and Knowledge Engineering*, Skokie, IL, Knowledge Systems Institute, pp. 284-293.

Fensel, D, 2004, "DERI Research Report," [Online]. Available:
http://www.deri.at/fileadmin/documents/DERI-TR-2004-05-31.pdf

Filman, R. E., 1998, "Achieving Ilities: Workshop on Compositional Software Architectures," Monterey, California. [Online]. Available:
http://www.objs.com/workshops/ws9801/papers/paper046.doc.

Fischer, G., A. Lemke, R. McCall, and A. Morch, 1995, "Making Argumentation Serve Design," in *Design Rationale Concepts, Techniques, and Use*, T. Moran and J. Carroll, (eds), Lawrence Erlbaum Associates, pp. 267-294.

Garlan, D., 1995a, "Design Techniques," *ACM SIGSOFT Software Engineering Notes*, July Vol. 20, No. 3, ACM Press, pp. 84 – 89.

Garlan, D, 1995b, "What is Style?" in *Proceeding of Dagsthul Workshop on Software Architecture*, [Online]. Available: http://citeseer.ist.psu.edu/garlan95what.html

Ghoniem, M., J.D. Fekete, and P. Castagliola, 2004, "A Comparison of the Readability of Graphs Using Node-Link and Matrix-Based Representations," *Proceedings of Information Visualization 2004*, Austin, Texas, pp. 17-24.

Gogolla, M., 1998, *UML for the Impatient*, Research Report 3/98, Universität Bremen, Germany.

Gomez-Perez, A. and V. R. Benjamins, 1999, "Overview of Knowledge Sharing and Re-use Components: Ontologies and Problem-Solving Methods," IJCAI-99 Workshop on Ontologies and Problem-Solving Methods (KRR5).Stockholm, Sweden, pp. 45-52.

Gomez-Perez, A., A. Moreno, J. Pazos, and A. Sierra-Alonso, 2000, Knowledge maps: An essential technique for conceptualization. *Data and Knowledge Engineering* Vol. 33, No. 4, pp. 169–190.

Grimes, G.J., 1997, "Photonic Packaging Using Laser/Receiver Arrays and Flexible Optical Circuits," *IEEE Transactions on Components, Packaging, and Manufacturing Technology (CPMT), Part B: Advanced Packaging*, Vol. 20, No. 4, pp. 33-37.

Grimes, G.J., 1994, "Packaging for Optical Backplanes," *IEEE Proceedings of the LEOS Annual Meeting*, pp. 222-223.

Grimes, G.J., 1995, "Optoelectronic Packaging for Optical Backplanes," *IEEE Proceedings of the 45th ECTC*, pp. 548-551.

Grimes, G.J., 2004, "The POF Value Proposition vs. Glass, Copper, and Wireless," Proceedings of the POF World Conference, San Jose, CA, pp. 24-27.

Grosso, W. E., H. Eriksson, R. W. Fergerson, J. H. Gennari, S. Tu, and M. A. Musen, 2000, " *Knowledge Modelling at the Millennium: The Design and Evolution of Protégé*, SMI Report SMI-1999-080, Stanford Medical Institute.

Gruber, T. R., 1993, *Toward principles for the design of ontologies used for knowledge sharing*, Report KSL-93-04, Stanford University.

Gruber, T. R., 1990, "Model-based Explanation of Design Rationale," in A Survey of Design Rationale Systems: Approaches, Representation, Capture and Retrieval, T. Gruber, *Engineering with Computers,* Vol. 16, No. 3-4, 2000, pp. 209-235.

GUARDS, 1997, "Approach, Methodology and Tools for Validation by Analytical Modelling," Technicatome/PDCC Second Part of D302: Functional Specification and Preliminary Design of GUARDS Validation Environment", by E. Jenn and M. Nelli, 1997, *ESPRIT Project,* 20716 GUARDS Report, February.

Guedez,V., P. Mondelo, A. Hernandez,and L. Mosquera, 2001, "Ergonomic design of small containers using the Quality Function Deployment (QFD)," International Conference on Computer-Aided Ergonomics and Safety. Maui, Hawaii. [Online].Available: http://cep.upc.es/Publicaciones/CAES2001/Ergonomicdesign.htm

Karsenty, L., 1996, "An Empirical Evaluation of Design Rationale Documents," in A Survey of Design Rationale Systems: Approaches, Representation, Capture and Retrieval, T. Gruber, *Engineering with Computers,* Vol. 16, No. 3-4, 2000, pp. 209-235.

Haimes, Y, 1981, "Risk-Benefit Analysis in a Multiobjective Framework," in *Risk/Benefit Analysis in Water Resources Planning and Management,* Y. Haimes (ed.) Plenum Press, New York.

HIDE, 1999, "High Level Integrated Design," [Online].Available: http://www3.informatik.uni-erlangen.de/Publications/Articles/dalcin_words99.pdf

Hoppe, T.1993," VVT Terminology: A Proposal," *IEEE Expert*, June, pp. 48-55.

Houlding, B., 2006, "On Sequential Decision Making with Adaptive Utilities," [Online].Available: http://www.dur.ac.uk/brett.houlding/resources/BH-FPAC.pdf

ICAD Release 7.0, 2004, "Knowledge based Engineering and the ICAD System," [Online]. Available: http://www.ktiworld.com/pdf/understanding-tis-2.pdf

IDOV, 2006, "Identify, Design, Optimize and Verify," [Online].Available: http://www.isixsigma.com/library/content/c020722a.asp

IEEE-1471, 2006, "AWG IEEE," [Online]. Available: http://standards.ieee.org/reading/ieee/std_public/description/se/1471-2000_desc.html

IHMC, 2006, "Florida Institute for Machine Cognition," [Online]. Available: http://www.coginst.uwf.edu/newsletters/IHMCnewslettervol3iss1.pdf

INCOSE, 2006a, "The International Council on Systems Engineering," [Online]. Available: http://www.incose.org/

INCOSE, 2006b, "Systems Engineering modeling language: SysML," [Online]. Available: http://www.sysml.org/

Internet2, 2004, [Online]. Available: http://whatis.techtarget.com/definition

ISO, 2002, [Online]. Available: http://www.15288.com/about_15288.htm

Jacobson, I., G. Booch, and J. Rumbaugh, 1999, *Unified Software Development Process*, Addison-Wesley, Reading, MA.

Jayaswal, B. K. and P. C. Patton, 2006, *Design for Trustworthy Software: Tools, Techniques, and Methodology of Developing Robust Software*, Prentice Hall, Upper Saddle River, NJ.

Jett, W., 2006, *An agent-based framework for software evolution*, PhD. Dissertation, UAB, Birmingham, AL.

Kirk, A. G., D. V. Plant, M. H. Ayliffe, M. Châteauneuf, and F. Lacroix, 2003,"Design Rules for Highly Parallel Free-Space Optical Interconnects," *IEEE journal of selected topics in quantum electronics*, Vol. 9, No. 2, pp. 231-245.

Klein, M., 1993, "DRCS: An Integrated System for Capture of Designs and Their Rationale," in *Artificial Intelligence in Design*, Gero, J. (ed.), Kluwer Academic Publishers, Norwell, MA, pp. 393-412.

Klein, M.: 1997, "An Exception Handling Approach to Enhancing Consistency, Completeness and Correctness in Collaborative Requirements Capture," in A Knowledge-based Approach to Handling Exceptions in Workflow Systems, 2000, *Computer Supported Cooperative Work*, Vol. 9, No. 3-4, 399-412.

Kotov V., 1997, "Systems of Systems as Communicating Structures," Computer Systems Laboratory, Technical Report, HPL-97-124 October, [Online]. Available: http://www.hpl.hp.com/techreports/97/HPL-97-124.pdf

Kruchten, P., 1999, *Rational Unified Process-An Introduction*, Addison-Wesley, Reading, MA.

Lambert, J. H., B. L. Schulte, and P. Sarda, 2005, "Tracking the complexity of interactions between risk incidents and engineering systems " *Systems Engineering*, Vol. 8, No. 3, pp. 262–277.

Larsen R. F., D. M. Buede, 2002, "Theoretical framework for the continuous early validation (CEaVa) method," *Systems Engineering*, Vol.5, No. 3, 223-241.

Lee, J., 1990, "SIBYL: A qualitative design management system." In *Artificial Intelligence at MIT: Expanding Frontiers*, P.H. Winston and S. Shellard (eds) MIT Press, Cambridge, MA, pp. 104-133.

Lee, J., 1997, Design Rationale Systems: Understanding the Issues, *IEEE Expert*, Vol. 12, No. 3, pp. 78-85.

Lee, S. H. and Y. C. Lee, 2006, "Optoelectronic Packaging for Optical Interconnects," *Optics & Photonics News,* Vol. 17, January, pp.40-45.

Lewis, E. E., 1987, *Introduction to Reliability Engineering*, John Wiley & Sons Inc., New York.

Liu, Y.S., 1997, "Progress in Optical Interconnects for Data Communication — Bringing Light to the Board, Backplane, and Intra-Boxes", Technical Report: Control Systems and Electronic Technology Laboratory-97CRD134.

Lowrance, W. W., 1976, Of *acceptable risk: Science and the Determination of Safety,* William Kaufman, Los Altos, CA.

Maciaszek, L. A., 2004, "Managing Complexity of Enterprise Information Systems," *ICEIS'2004 Sixth Int. Conf. on Enterprise Information Systems*, Volume I, INSTICC, Portugal, pp. 17-23.

Maciaszek, L. A., 2001, *Requirements Analysis and System Design, Developing Information Systems with UML*, Addison-Wesley, Reading, MA.

Madni, A.M.,2006, "System-of-Systems Architecting: Critical Success Factors,"[Online]. Available:http://sunset.usc.edu/events/2006/CSSE_Convocation/presentations/Madni.ppt

Madni, A.M., W. Lin, and C. C. Madni, 2001, "IDEON™: An Extensible Ontology for Designing, Integrating, and Managing Collaborative Distributed Enterprises" *Systems Engineering*, Vol. 4, No. 1, [Online]. Available: http://ieeexplore.ieee.org/iel4/5875/15661/00725050.pdf?arnumber=725050

Madni, A.M., A. Sage, and C.C. Madni, 2005, "Infusion of Cognitive Engineering into Systems Engineering Processes and Practices," *IEEE International Conference on Systems, Man, and Cybernetics,* Vol.1, pp.960-965.

Mahafza, R. S., P. Componation, and D. Tippet, 2005, "A performance-based technology assessment methodology to support DoD acquisition," [Online]. Available: http://www.dau.mil/pubs/arq/2005arq/2005arq-37/mahafza.pdf

Maier M. and E. Rechtin, 2000, *The Art of Systems Architecting,* 2nd Edition, CRC Press, Boca Raton, FL.

Manjarres, A., S. Pickin, and J. Mira, 2002, "Knowledge model reuse: therapy decision through specialization of a generic decision model," *Expert Systems with Applications,* Vol. 23, No. 2, pp. 113-135.

Maclean, A, Young, R., Bellotti, V. and Moran, T., 1991, "Design space analysis: bridging from theory to practice via design rationale," Xerox EuroPARC, [Online]. Available: http://citeseer.ist.psu.edu/cache/papers/cs/3445/http:zSzzSzwww.rxrc.xerox.comzSzpubli szSzcam-trszSzhtmlzSz..zSzpszSz1991zSzepc-1991-128.pdf/maclean91design.pdf

Milton, N., 2004, "Esprit Moka Modeling Language," [Online]. Available: http://www.epistemics.co.uk/Notes/146-0-0.htm

Miller, E., 2004, "Official W3C Semantic Web page," [Online]. Available: http://www.w3.org/2001/sw

Moka, 2004, "KBE MOKA Modeling Language," [Online]. Available: http://www.epistemics.co.uk/Notes/146-0-0.htm

Muirhead, B.K., 1999, "Deep space 4/Champollion, 2nd generation cheaper, better, faster,"Aerospace Conference, Pasadena, CA., [Online]. Available: http://ieeexplore.ieee.org/iel5/6411/17211/00793136.pdf

Muirhead, B.K., 2004,"Mars Rovers, Past and Future,"Aerospace Conference, Pasadena, CA., [Online]. Available: http://ieeexplore.ieee.org/iel5/9422/29900/01367598.pdf

Muirhead, B.K., W. L. Simon,. 1999, *High Velocity Leadership: The Mars Pathfinder Approach to Faster, Better, Cheaper*, Harper Collins, New York.

Myers, K., N. Zumel, and P. Garcia, 1999, "Automated Capture of Rationale for the Detailed Design Process," in *Proceedings of the Eleventh National Conference on Innovative Applications of Artificial Intelligence,* AAAI Press, Menlo Park, CA, pp. 876-883.

NASA GSFC, 1998, "Software Metrics and Reliability," [Online]. Available: http://satc.gsfc.nasa.gov/support/ISSRE_NOV98/software_metrics_and_reliability.html

NASA, 2000a, "NASA Scientific and Technical Information Program, Probabilistic Risk Assessment - A Bibliography," *NASA/SP-2000-6112,* July, pp. 83. [Online]. Available: http://www.tpub.com/content/nasa2000/

NASA, 2000b, "Probabilistic Risk Assessment: What Is It And Why Is It Worth Performing It?" [Online]. Available: http://www.hq.nasa.gov/office/codeq/qnews/pra.pdf

NASA, 2002, "Risk Management," [Online]. Available: http://www.hq.nasa.gov/office/codeq/risk/risk.htm

NASA JSC, 2005, "Public Lessons Learned Entry: 0825," [Online]. Available: http://www.nasa.gov/offices/oce/llis/0825.html

NRC, 2006, "National Research Council," [Online]. Available: http://www.nationalacademies.org/nrc/

NRC, 2002, *Approaches to Improve Engineering Design*, National Academies Press, [Online]. Available: http://www.nap.edu/catalog/10502.html

NGI, 2006, "Next Generation Internet," [Online]. Available: http://europa.eu.int/information_society/policy/nextweb/grid/index_en.htm

Nyquist, J.S., Sherman, C.J., and Grimes, G.J., 2000, "Systems Level Packaging of High Density Optoelectronic Interconnections," *Proceedings of the 50th Electronic Components and Technology Conference (ECTC)*, pp. 1272-1277.

Oldham, K., Kneebone, S., Callot, M., Murton, A. and Brimble, R., in Mårtensson, N., Mackay, R., and Björgvinsson, S., (eds.), 1998, "MOKA: Methodology and tools Oriented to Knowledge based engineering Applications: Changing the Ways We Work, Advances in Design and Manufacturing," *Proceedings of the Conference on Integration in Manufacturing*, Volume 8, IOS Press, Amsterdam, pp. 198-207.

OMG, 2005, "UML 2.0," [Online]. Available: http://www.omg.org/gettingstarted/what_is_uml.htm

OMG, 2006, "Model-Driven-Architecture," [Online]. Available: http://www.omg.org/mda/

Ontology, 2006, "Laboratory for Applied Ontology," [Online]. Available: http://www.loa-cnr.it/

Pahl, G. and Beitz, W., 1988, *Engineering Design: A Systematic Approach*, K.Wallace, (ed.), Springer-Verlag, New York.

Peña-Mora, F. and S. Vadhavkar, 1996, "Augmenting design patterns with design rationale," in *Artificial Intelligence for Engineering Design, Analysis and Manufacturing*, Vol. 11, Cambridge University Press, West Nyack, New York, pp. 93-108.

Peña-Mora, F., Sriram, D. and Logcher, R. 1995, Design Rationale for Computer - Supported Conflict Mitigation, *ASCE Journal of Computing in Civil Engineering*, pp. 57-72.

Perera, J., 2002, "An Integrated Risk Management Tool and Process," NASA-Johnson Space Center Houston, Texas, [Online]. Available: http://ieeexplore.ieee.org/iel5/10432/33126/01559306.pdf?arnumber=1559306

Pimmler, T. U. and S. D. Eppinger, 1994, "Integration Analysis of Product Decompositions." *Proc of ASME Design Engineering Technical Conferences*. Minneapolis, MN. pp. 55- 62.

Powell, D., J. Arlat, L. Beus-Dukic, A. Bondavalli, P. Coppola, A. Fantechi, E. Jenn, C. Rabejac, A. Wellings, 1999, "GUARDS: a generic upgradable architecture for real-time dependablesystems," *IEEE Transactions on Parallel and Distributed Systems*, Vol. 10, No. 6, pp. 580-599

PRA, 2006, "Probabilistic Risk Assessment," [Online]. Available: http://ntrs.nasa.gov/search.jsp?No=30&Ne=35&N=4294885333

Reynolds, D., 2005, "Jena 2 inference support," [Online]. Available: http://jena.sourceforge.net/inference/

Ring, J. and A. M. Madni, 2005, "Key Challenges and Opportunities in 'System of Systems' Engineering," *IEEE International Conference on Systems, Man, and Cybernetics*, Vol. 1, pp. 973-978.

Robertson, P. J., Chen, H. Y., Brandt, J. M., Sullivan, C. T., Pierson, L.G., Witzke, E. L., Gass, K., 2000, *Optical Backplane/Interconnect for High Speed Communication LDRD*, Technical Report, SAND2001-0684 Sandia National Laboratories, New Mexico.

Robinson, P., and F. Gout, 2004, "Extreme Architecture, A Minimalist IT Architecture Framework," *The UK Oracle User Group Journal*, Vol. 19, pp. 34-37

Ross, D. T., 1985, "Applications and Extensions of SADT," *IEEE Computer*, April, pp. 25-34.

Russel, S. and P. Norvig, 2003, *Artificial Intelligence A Modern Approach*, Pearson Education, Prentice-Hall, Upper Saddle River, NJ.

Sage A. P. and C. D. Cuppan, 2001, "On the Systems Engineering and Management of Systems of Systems and Federations of Systems," *Information, Knowledge, Systems Management*, Vol. 2, No. 4, pp. 325-45.

Sage, A. P., and J. D. Palmer, J.D., 1990, *Software Systems Engineering*, John Wiley & Sons Inc., New York.

Sandia National Laboratories, 2006, "Java Expert System (JESS)," [Online]. Available: http://herzberg.ca.sandia.gov/jess/

Satyanarayanan, M., J.J. Kistler, P. Kumar, M.E. Okasaki, E.H. Siegel, and D.C. Steere, 1990, "Coda: A Highly Available File System for a Distributed Workstation Environment," *IEEE Transactions on Computers*, vol. 39, no. 4, pp. 447-459, [Online]. Available: http://citeseer.ist.psu.edu/satyanarayanan90coda.html

Schneidewind, N. F., 2002, "Report on Results of Discriminate Analysis Experiment," *27th Annual NASA Goddard/IEEE Software Engineering Workshop Proceedings*, Naval Postgraduate School., Monterey, CA, pp. 9-16.

Schreiber, G., H. Akkermans, A. Anjewierden, R. de Hoog, N. Shadbolt, W. Van de Velde, and B. Wielinga, 2000, *Knowledge Engineering and Management, The Common-KADS Methodology*, MIT Press, Cambridge, MA.

Schulmeyer, G. G. and G. R. MacKenzie, 1999, *Verification & Validation of Modern Software-Intensive Systems*, Prentice Hall, Upper Saddle River, NJ.

Seker, R. and. Tanik, M.M, 2004, "An Information-Theoretical Framework for Modeling Component-based systems," *IEEE, SMS, Part C: Applications and Reviews*, Vol. 34, No. 4, pp. 475-485.

Sewell, M., and L. Sewell, 2002, *The Software Architect's Profession: An Introduction*, Prentice Hall, Upper Saddle River, NJ.

Shannon, C.E., 1948, "A Mathematical Theory of Communication," *Bell System Technical Journal*, 27, pp. 379–423.

Shaw, M., D. Garlan, 1995, "Formulations and Formalisms in Software Architecture," in *Computer Science Today: Recent Trends and Developments*, Lecture Notes in Computer Science, Volume 1000, Springer-Verlag, New York, pp. 307-323.

Shaw M. and D. Garlan, 1996, *Software Architecture: Perspectives on an Emerging Discipline*, Prentice-Hall, Upper Saddle River, NJ.

Shipman, F. and R. McCall, 1996, "Integrating different perspectives on design rationale: Supporting the emergence of design rationale from design communication," in *Artificial Intelligence for Engineering Design, Analysis, and Manufacturing*, Vol. 11, Cambridge University Press, West Nyack, New York, pp. 141-154.

Simon, H. A., 1981, *The Sciences of the Artificial*, MIT Press, Cambridge, MA.

Sloane, N. J. A. and A. D. Wyner (eds.), 1993, *Claude Elwood Shannon: collected papers*, IEEE Press, New York.

Speel, P., A. T. Schreiber, W. van Joolingen, and G. Beijer, 2001, "Conceptual Models for Knowledge-Based Systems," in Avram Gabriela (2005) "Empirical Study on Knowledge Based Systems" The Electronic Journal of Information Systems Evaluation, Vol. 8, No.1, pp 11-20, [Online]. Available: http://ejise.com/volume-8/v8-iss-1/v8-i1-art2-avram.pdf

The Standish Group International, 1994, "The chaos Report," [Online]. Available: http://cs.hbg.psu.edu/comp413/chaos1994.pdf

Stokes, M., 2001, "Managing Engineering Knowledge: MOKA - Methodology for Knowledge Based Engineering Applications," [Online]. Available: http://www.epistemics.co.uk/Notes/146-0-0.htm
Studer, R., R. V. Benjamins, and D. Fensel, 1998, "Knowledge Engineering: Principles and Methods," *IEEE Data & Knowledge Engineering*," Vol. 25, pp. 161-197.

Suh, N., 2001, *Axiomatic Design, Advances and Applications,* Oxford University Press, New York.

Tanik, U.J., G. J. Grimes, C.. J. Sherman , V. P. Gurupur, 2005, "An Intelligent Design Framework for Optical Backplane Engineering," *Journal of Integrated Design and Process Science,* Vol. 9, No. 1, pp.41-53.

Telelogic, 2006, [Online]. Available: http://www.telelogic.com/corp/index.cfm .

Thome, B. (Ed.), 1993, *Systems Engineering: Principles and Practice of Computer-Based Systems Engineering*, John Wiley & Sons Inc., New York.

Trevino, L., D. E. Paris, and M. Watson, 2005, *A Framework for Integration of IVHM Technologies for Intelligent Integration for Vehicle Management*, NASA Marshall Space Flight Center, Technical Report, Document ID: 20050162249, NTRS: 2005-05-17.

Trewn, J. and K. Yang, 2000, "A Treatise on System Reliability And Design Complexity," *First International Conference on Axiomatic Design, Proceedings of ICAD-2000,* Cambridge, MA, pp. 162-168.

TRIZ, 2006, [Online]. Available: www.triz40.com

Ulrich, K. T. and S. D. Eppinger, 2004, *Product Design and Development,* 3rd Edition, McGraw-Hill, New York.

Valishevsky, A. 2003, "Using granular-evidence-based adaptive networks for sensitivity analysis," [Online]. Available: http://citeseer.ist.psu.edu/cache/papers/cs/26821/http:zSzzSzhome.lanet.lvzSz~md80022zSzangie-case.pdf/valishevsky02using.pdf

Valishevsky A. and A. Borisov, 2003, "Using interval-valued entropy for risk assessment," Proceedings of the International Conference Modeling and Simulation of Business Systems, pp. 74-78. [Online]. Available: http://citeseer.ist.psu.edu/cache/papers/cs/27095/http:zSzzSzhome.lanet.lvzSz~md80022zSzentropy.pdf/valishevsky03using.pdf

OWL, 2006, "Visual Web Ontology Language," [Online]. Available:
http://www.visualmodeling.com/VisualOWL.htm

W3C, 2006, "The World Wide Web Consortium," [Online]. Avaiable:
http://www.w3.org/

W3C, 2006, "Semantic Web," [Online]. Available: http://www.w3.org/2001/sw/
Wang L., W. Shen, H. Xie, J. Neelamkavil, and A. Pardasani, 2001, "Collaborative Conceptual Design – State of the Art and Future Trends," *Computer Aided Design*, Vol. 34, No. 13, pp. 981-996.

Wortman, B., 2001, "Quality Council of Indiana," [Online]. Available:
http://www.qualitycouncil.com/samples/CSSBB_2002_001_Instructor_Sample.pdf

Zozayza-Gorostiza, C. and Hendrickson, C., 1987, "An Expert System for Traffic Signal Setting Assistance," *ASCE Journal of Transportation Engineering*, Vol. 113, No. 2, pp. 108-126.

Zrymiak, D., 2006, "Software Quality Function Deployment Modifying the House Of Quality for Software," [Online]. Available:
http://software.isixsigma.com/library/content/c030709a.asp

APPENDIX

A SOFTWARE ARCHITECTURE VALIDATION DETAIL

Validation Division I

(I) SOFTWARE ARCHITECTURE

Validation Area I -A

(A) Architecture development process

Methodology: Applying 16 standards, techniques, methods

I-A-1

(1) ANSI/IEEE Std. 1471 terminology

IEEE 1471 is a short name for a standard known as *ANSI/IEEE Std.1471-2000,*

Recommended Practice for Architecture Description of Software-Intensive Systems,

recently accepted by ISO JTC1 as ISO/IEC DIS 25961.   It is the formal standard for

software or system architecture, providing definitions, such as view and viewpoint, and

other key items to form the basis for the terminologies used by the AIDF [IEEE-1471,

2006; Ayliffe, M.H. 2001; INCOSE, 2006a; NRC, 2006].

I-A-2

(2) National Defense Authorisation Act for Fiscal Year 1996 (Clinger-Cohen Act)

The Clinger-Cohen Act, also known as the Information Technology Management Reform Act, is a United States federal law that was co-authored by U.S. Representative William Clinger and Senator William Cohen in 1996. Intended to improve the way the federal government acquires and manages information technology (IT), the law mandates that all Federal agencies use performance-based management principles for acquiring IT. An example architecture framework is DoDAF used by the military that evolved from the earlier Zachman framework.

I-A-3

(3) OMG-Model-Driven Architecture (MDA) technology independence approach
The Model-Driven Architecture is a software design approach that was officially launched in 2001 by its sponsor, the Object Management Group (OMG). MDA is intended to support model-driven engineering of software systems. The MDA is a specification providing a set of guidelines for structuring specifications expressed as models. System functionality may first be defined as a platform-independent model (PIM), with the goal of separating design from architecture, so that the design, represented by the functional requirements, can last longer than the architecture and ensuing technologies developed to realize it for long-term design survivability.

I-A-4

(4) First International Workshop on IT Architectures for Software Systems in 1995
This workshop held in 1995 was an important landmark in software development. As a result of this conference, Shaw, representing Carnegie Mellon, was able to distill and

categorize the submitted papers into five categories defining architecture types and styles, one of which is framework using pipes and filters – used by the AIDF.

I-A-5

(5) Shaw-Carnegie Mellon architectural categorizations

After the First International Workshop on IT Architectures for Software Systems in 1995, Shaw categorized architecture into five categories: framework model, structural model, dynamic model, process model, and functional model. The architectural framework mode of the AIDF is comprised of the architecture components in the form of DP, as in the engine block represented as modules. The structural and dynamic models are developed after the framework is completed and set to software engineering for further OO development in UML. The process model is developed using the axiomatic V-Model, enabling outsourcing of individual FR to global software teams for subsequent integration by the AIDF. The functional model of the AIDF is represented solely by the functional requirements.

I-A-6

(6) ATAM for architectural trade-off analysis

ATAM stands for architecture trade-off analysis model, which enables the development of robust architecture for the AIDF by using various methods to seek trade-offs in architecture design for achieving different functionality. ATAM is usually used in conjunction with CBAM.

I-A-7

(7) CBAM for architectural cost assessment

CBAM stands for cost benefit architecture model, which enables the development of cost-effective architecture by attending to cost parameters associated for the design.

I-A-8

(8) Acclaro Design for Six Sigma (DFSS) implementation

Axiomatic Design for Six Sigma is an architectural development tool provided by Axiomatic Design solutions, Inc. Also acquired by General Dynamics for architecture development, we used Acclaro DFSS for the implementation of the architectural framework, as it captures the functional requirements and design parameters essential for a MDA approach. Its DFSS functionality provides a way to capture information in a format to apply DFSS IDOV front-end architecture validation (Pre-deployment).

I-A-9

(9) Telelogic TAU implementation of UML and SysML

Telelogic TAU provides a full-scale development environment to construct OO diagrams for structural modeling of the AIDF modules. Although outside the scope of the thesis design framework, it is useful to introduce this tool, so that software developers can see how the AIDF output, whose hundreds of FR segments can be mapped to software engineering methods for further OO model development independent of other segmented FRs in the hierarchically decomposed framework, as in the case of detailing module element interactions, followed by actual coding.

I-A-10

(10) Rational Unified Process (RUP)

The Rational Unified Process is important to mention in this thesis because this is a popular method in industry for architecture management to follow that is consistent with the framework model development.

I-A-11

(11) Architecture-driven software construction

Architecture-driven software construction is critical to understanding the steps taken to construct the AIDF. Most importantly, it clearly separates the architecting process from the software engineering process, which is considered in the build phase, outside the scope of this thesis.

I-A-12

(12) Waterfall Model

The waterfall method is a popular method, although it has had its detractors, including the originator of the method, Royce. He and others contend that it does not necessarily take into account iterative development, although the AIDF does introduce iteration at the earliest stage of engineering development, during the concept formation stage. Thus, we spend ample time on developing accurate functional requirements representing the interests of the end-user through rapid prototyping with Acclaro DFSS and QFD, saving on prototyping cost down the line – the iteration that oftentimes leads to project frustration or failure.

I-A-13

(13) Spiral Model

The spiral model is also a popular model, where development is done in cycles with risk mitigation introduced to each prototype, introduced by Boehm. We selected this model as a potential development paradigm for the AIDF, since it emphasized iteration of risk mitigation.

I-A-14

(14) Capability Maturity Model Integration (CMMI)

CMMI is a process improvement method usually utilized by companies for appraising their processes according to a standard developed by the Software Engineering Institute of Carnegie Mellon. Although we did not need to use CMMI directly in the development of the AIDF, its processes were reviewed.

I-A-15

(15) Rapid Application Development (RAD)

RAD is an approach to prototype development in a way that emphasizes speed, as the name denotes. We were able to utilize the concepts of RAD, after we acquired the archi-tecture development tool, Acclaro DFSS, which allowed us to experiment with various representative designs having different structures for FR and DP composition of the AIDF. After some trial and error, we were able to realize the current AIDF structure e m-phasizing modularity for a high-degree-of granularity and resolution. This enabled us to think more clearly on the configuration of the engine, by eliminating any unnecessary clutter and focusing directly on interlacing the engine blocks defined in the KCE.

I-A-16

(16) Dynamic Systems Development Method

Part of RAD, this approach fixes time and resources for p roducing a viable prototype. We found the techniques of this approach intriguing considering that we wished to develop many different prototypes with limited resources and time at the graduate student level.

Validation Area I –B

(B) Architecture framework model

Methodology: applying 7 standards, techniques, methods

I-B-1

(1)   AIDF implementation/Acclaro Design for Six Sigma (DFSS)

In order to develop the architecture framework mode, we were able to develop the fun c-tional model using FR and the framework model with FR/DP decomposition, where the DP were represented using the filter/pipe style. Subsequently we were able to implement the framework based on MDA approach using Acclaro DFSS, successfully reflecting the three stages of design support for each design p hase.

I-B-2

(2)   Failure Mode and Effects Analysis (FMEA)

FMEA is a method by which component hazards can be identified and their potential solutions developed before any problems actually do arise. This approach is excellent as a
basis for automating diagnosis methods. We introduce this method as a widely-
recognized and established industrial approach to trouble-shooting, and therefore we have
introduced this into the engine block as a viable design method for automation.

I-B-3

(3)   Design Structure Matrix (DSM) dependency resolution

DSM is a method by which components can be systematically compared to each prima r-
ily with a numerical schema, ideal for machine processing and easy web-enabled input by
domain experts. Due to its wide use in engineering as a method for easily viewing information, energy, and materials exchange, in addition to its knowledge engineering advantages we selected this as another module for design automation.

I-B-4

(4) Axiomatic Design Theory (ADT) risk mitigation

Based on its relative novelty in industry, with ample endorsement by government, we selected this axiomatic design theory as the bedrock for knowledge engineering of a rule-
base, especially since it was also developed at Massachusetts Institute of Technology and
supported commercially by Axiomatic Design, Inc. Furthermore, axiomatic design theory has received the highest rating in the document *Approaches to Engineering Design,*

published in 2001 and endorsed by the National Research Council, assessing the best design practices in the nation. Axiomatic design has two axioms, which has associated with it many related corollaries and theorems that are ideal for structuring a meta-level rule-base dealing only with FR and DPs, especially useful in conceptual design stage automation. The second axiom, information axiom, is especially useful in reducing information content, thereby optimizing the design for risk mitigation.

I-B-5

(5) Quality Function Deployment (QFD)

QFD has many applications, and when combined with an ISO standard for IT developed in 1991, we have an excellent basis to develop engineering features, in terms of requirements specifications, that clearly reflect end-user needs for purposes of validation and verification of architecture before implementation begins after framework development ends and software engineering and coding begins. For purposes of design automation, QFD is ideal for capturing tacit and explicit knowledge through a well-defined and popular template that can be made available online for ergonomic knowledge engineering with domain experts. The numerical aspects of the House of Quality can be easily converted into numerical ratings that are useful in weighting of rules and knowledge processing by intelligent agents.

I-B-6

(6) Generic Architecture for Upgradeable Real-Time Dependable Systems (GAURDS)
validation

The GAURDS validation framework is very suitable for an architectural framework for a KBE SoS application in the field of optical backplane engineering, especially because it us upgradeable and emphasizes real-time systems. GAURDS appears to be a relatively new development fulfilling the need for industry-grade architecture validation used by safety-critical systems, including space systems, railways, and nuclear submarines. For the purpose of ensuring high-throughput and reliability for the case study application of OBIT, this architecture appears ideal for post-deployment, as new COTS components are added and validated individually, as the system evolves.

I-B-7

(7) National Academy for Engineering state-of-the-art 2001 report outlining Approaches to Engineering Design

The National Academy for Engineering *Approaches to Engineering Design* report was a critical find during the research process, as it substantiated the claims of axiomatic design by an independent government report, stating best-practices in 2001. Axiomatic design received the highest rating in the category assisting in alternative designs, and the Academy report was endorsed by NRC.

Validation Area I-C

Architectural Structural/Dynamic Model

Methodology: applying 3 standards, techniques, methods

I-C-1

(1) High Level Integrated Design Environment (HIDE) for dependability

HIDE is an approach to validate the structural model done primarily in objected oriented languages, such as UML. We recommend this validation process for software engineering based on the AIDF framework.

I-C-2

(2) Telelogic TAU UML/SysML implementation

We acquired Telelogic TAU from the vendor during the research process to test to see if the structural models can be developed using this tool for large-scale development. We confirmed that in addition to UML development, the tool came equipped with SysML as well, allowing for systems engineering specification using the recently accepted SysML standard, April 2006. As part of the testing process, we joined a SysML user group that is frequented by members of INCOSE.

I-C-3

(3) NASA SATC metrics for object-oriented code evaluation in terms of classes, methods, cohesion, and coupling

The NASA SATC metrics are useful for software development and coding after the initial functional and framework modeling, followed by structural and dynamic modeling, is complete. These provide a methodology to help optimize code development, when combined with architecture-driven software development of the AIDF.

Validation Area I-D

Process Model

Methodology: applying 2 standards, techniques, methods

I-D-1

(1) Axiomatic V-Model mapping process

The V-model process, based on the NRC endorsed axiomatic design theory, provides a methodology to map the framework model to the structural/dynamic model, providing a means for the benefits of architecture-driven process to seamlessly map to object-oriented development of modules. First, a hierarchically decomposed FR/DP design matrix can be constructed, where hundreds of FR can be defined and mapped directly to multiple DP fulfilling that requirement independent of other FR. Then the DP, representing design or inference engine modules and elements in the engine block, for instance, can be configured and outsourced as individual strips of FR, treated as methods to be developed by software engineering teams. Subsequently, the AIDF allows a single architect to achieve intellectual control over a large-scale project by providing a means to integrate the developed methods into a full-scale software artifact, such as a configured KBE SoS for a particular automated design support application, such as optical backplane engineering.

I-D-2

(2) National Academy of Engineering

The process model we use in the AIDF is based on the V-model of axiomatic design theory, which is highly rated and promoted as a viable engineering design approach in the report of the National Academy of Engineering. The United States National Academy of

Engineering is a private, non-profit institution, founded in 1964. Providing oversight of certain key reports such as best practices listed in *Approaches to Engineering Design* in 2001, the National Research Council of the United States was organized in 1916 and is the operational arm of the United States National Academy of Sciences and the United States National Academy of Engineering.

B VENDORS SUPPLYING TECHNOLOGIES FOR VALIDATION

GAURDS Validation Framework applied after front-end Six Sigma Validation

A consortium of European companies and academic partners was formed to design and develop a Generic Upgradeable Architecture for Real-time Dependable Systems
(GUARDS), together with an associated development and validation environment. The
strength of GAURDS is that is has been developed, promoted, tested, and accepted as an
industrial-grade validation framework for any instance of a functional architecture, such
as the AIDF [HIDE, 1999; ICAD Release 7.0, 2004; IDOV, 2006; NRC, 2006]. Government support was provided by the European Union for GAURDS through the ESPRIT
Project 20716 initiative [Powell et al. 1999]. In order to validate the AIDF architectural
framework, we have selected the GAURDS approach as a comprehensive validation
methodology for functional architecture for real-time KBE systems, after front-end validation is completed with Acclaro Design for Six Sigma. Further more, since GAURDS
needs to have a functional architecture to validate, the AIDF architectural framework implementation using Acclaro DFSS is ideal because it breaks down the requirements hierarchically according to functional requirements

Example Applications of GAURDS

In order to be considered generic, the architecture must be able to meet the widest
possible spectrum of dependability and real-time requirements, especially meeting standards ensuring extremely low catastrophic failure rates for individual subsystems.
Some examples of GAURDS application has been in the areas of architecting sensitive,
mission-critical systems, specifically in the domain of nuclear submarines, space systems,

and railways. We will apply GAURDS validation framework to architecting a KBE system using the AIDF architectural framework in the domain of OBIT systems, also requiring low catastrophic failure rates. Thus, the GAURDS approach to validation of the AIDF provides a state-of-the-art methodology for ensuring that the AIDF model's functional architecture is valid. This step is crucial, considering that all the other models - starting with the structural model - depend on this meta-level validation step.

Rationale for Utilizing GAURDS as a Validation Framework for the AIDF

The goal of GAURDS is to be able to provide a method to configure instances of a generic architecture that can be shown to meet the diverse requirements of critical real-time application domains, such as AI design support for product engineering applications. The cost of validation and certification of instances of the architecture, when customized to the application of the AIDF, is a critical factor which is also treated by this approach. There are advantages of the GAURDS validation framework, when applied to the AIDF functional architecture. GAURDS fits well with the axiomatic FR/DP hierarchical decomposition process of the AIDF; it enables the re-use of already-validated components in different instances, making the AIDF amenable to multiple domains for broad impact; it concentrates validation obligations on a minimum set of critical components, allowing focused KBE validation; it encourages the use of Commercial-off-the-shelf (COTS) components, allowing the modular development of the AIDF inference (Table B-1) and design (Table B-2) engines. The vendors supplying technologies to the modules can be depicted in a color-coded diagram (Fig. B-1).

Table B-1. Sample baseline tools to develop automated inference capability.

AIDF Inference Engine Module	Sample Vendor Web Access Date: 11-7-06	Sample baseline tool to develop automated, networked capability	Related Features
(1) domain rule support	Infoharvest http://www.infohar vest.com	Criterium DecisionPlus 3.0 (CDP 3.0)	Resource provided with Systems Engineering Best Practices Center
(2) predicate logic support	Sandia National Laboratories http://www.cs.vu.nl /~ksprac/2002/doc/ Jess60/intro.html	Java Expert System Shell (JESS) Jess 6.0	Supports the development of rule-based expert systems; compatible with all versions of Java
(3) algorithmic reasoning support	Mathworks http://www.mathw orks.com/products/ matlab/	MATLAB 7.3	Provides high-level language/multi-purpose algorithmic development tools; arithmetic operators, flow control, data structures, data types, object-oriented programming (OOP), and debugging features
(4) fuzzy logic support	Mathworks http://www.mathw orks.com/products/ fuzzylogic/	MATLAB Fuzzy Logic Toolbox	Fuzzy clustering and adaptive neurofuzzy learning
(5) neural network support	Mathworks http://www.mathw orks.com/products/ neuralnet/	MATLAB Neural Network Toolbox	Provides pattern recognition, nonlinear system identification and control
(6) genetic algorithm support	Mathworks http://www.mathw orks.com/products/ gads/	MATLAB Genetic Algorithm Toolbox	Provides for initial population, fitness scaling, crossover, mutation
(7) Conant transmission support	National Technology Transfer Center; NASA case number MFS-31115	Space Debris Surfaces: Structural Damage Prediction and Analysis (SD SURF)	Provides transmission analysis to detect component influence
(8) Calibrated Bayesian Support	Bayesia, Ltd http://www.bayesia .com/	BayesiaLab	Inference by Junction tree, Gibbs sampling
(9) Data Mining Support	Data-Miner Pty Ltd http://www.data-miner.com/	The Rule Induction Kit (RIK)	Complete software package for discovering highly compact decision rules from data

Table B-2. Sample baseline tools to develop automated design capability.

AIDF Design Engine Module	Vendor Web Access Date: 11-7-06	Sample baseline tool to develop automated, networked capability	Related Features
(1) Axiomatic Design Theory (ADT)	HP Labs http://jena.sourceforge.net /	Java framework for building Semantic Web applications	Reading and writing RDF in RDF/XML, N3 and N-Triples/OWL; SPARQL query engine
(2) Theory of Inventive Problem Solving (TRIZ)	AxiomaticDesign, Inc, http://www.axiomaticdesi gn.com/	Acclaro Design for Six Sigma (DFSS)/TRIZ feature	Contradiction Matrix, provides design matrix support with improving/worsening feature comparison
(3) Hierarchical Multi-layer Design (MLH)	National Technology Transfer Center; NASA case number MFS-28573 http://www.openchannels oft-ware.com/projects/CASE_A	Computer Aided (Case-A) System Engineering and Analysis, ECLSS/ATCS Series	Schematic management system allows arrange icons representing system components
(4) Quality Function Deployment	Relex http://www.relex.com/pro ducts/humanfactors.asp	Relex Reliability Studio/ Human Factors Risk Analysis software	Verification/validation; Alarms, alerts, warnings
(5) Design Structure Matrix (DSM)	National Technology Transfer Center; NASA case number LAR-14793 http://www.openchannels oft-ware.com/projects/DEMA ID/&id=ntb	DeMAID (A Design Manager's Aid for Intelligent Decomposition);	Knowledge-based system for ordering the sequence of modules and identifying a possible multilevel structure for the design problem
(6) Fault-Tree-Analysis (FTA)	IsographDirect http://www.isograph.com/	Fault-tree+ Version 6	Minimal Cut Set calculation engine and gate types, such as AND, OR, NOR, NAND, NOT, XOR, Transfer, Voting, Inhibit
(7) Failure-Mode and Effects Analysis	Relex http://www.relex.com/pro ducts/fmeafmeca.asp	Relex Reliability Studio/FMEA/FMECA	Risk priority numbers (RPN), criticality ranks, risk levels, criticality matrices
(8) Reliability Block Diagram (RBD) Analysis	ReliaSoft http://blocksim.reliasoft.c om/features.htm	BlockSim 6 standard	Discrete event simulation engine/ complete analysis of highly complex configurable systems
(9)Technology Risk Factor (TRF) Assessment	Relex http://www.relex.com/pro ducts/relpredsoft.asp	Relex Reliability Studio/® Reliability Prediction software	MTBF/PRISM Process/ Telcordia Methods
(10) Entropy (ETP) Analysis	Relex http://www.relex.com/pro ducts/fracas.asp	Relex Reliability Studio/Failure Reporting, Analysis, and Corrective Action Systems (FRA-CAS)	Data Reporting, Analysis, and Corrective Action System
(11) Case Study: Optical Backplane Engineering (OPT) Domain	Mathworks http://www.mathworks.co m/products/optimization/	MATLAB Optimization Toolbox	Nonlinear optimization, nonlinear least squares

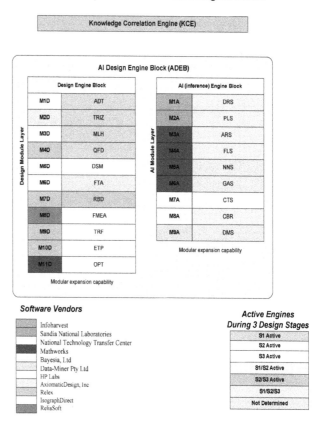

Software vendors supplying technology for modular development of AIDF dual engine block

Fig. B-1. Vendors supplying technology for dual engine block.